Underground

THE CITY IN THE TWENTY-FIRST CENTURY

Eugenie L. Birch and Susan M. Wachter, Series Editors

A complete list of books in the series is available from
the publisher.

UNDERGROUND

Dreams and Degradations in Bucharest

Bruce O'Neill

PENN

UNIVERSITY OF PENNSYLVANIA PRESS

PHILADELPHIA

Published by
University of Pennsylvania Press
Philadelphia, Pennsylvania 19104-4112
www.upenn.edu/pennpress

Printed in the United States of America on acid-free paper

10 9 8 7 6 5 4 3 2 1

Hardcover ISBN: 978-1-5128-2582-4
Paperback ISBN: 978-1-5128-2583-1
eBook ISBN: 978-1-5128-2584-8

A catalogue record for this book is available from the
Library of Congress.

To Rosemary and Leo

CONTENTS

PREFACE

While conducting the fieldwork for my first book about people experiencing homelessness in Bucharest between 2010 and 2011 (*The Space of Boredom*), I became aware of a longstanding encampment of so-called "street children" that had formed in the tunnels and canals beneath Bucharest's main railway station, the Gara de Nord. I also learned that unhoused people were squatting in the basements of otherwise vacant homes nearby. I began tracking how both groups relied upon the Gara de Nord's Metro station as a place to hang out, to earn and, ultimately, to spend money by working odd jobs for the Metro station's low-cost kiosks ("The Ethnographic Negative"). I initially imagined *Underground* as a study of this articulation of the city's subterranean extensions. My initial intention was to understand how, for these unhoused persons, urban life extended downward beneath the city's sidewalks.

While this study's underlying questions endured, nothing about the research unfolded as I had initially imagined. As became clear from the moment that I directed my full attention to this project, the underground is unsettled terrain. In the name of health and safety, the police forcibly closed the encampment beneath the Gara de Nord in the summer of 2015. At the same time, the close contacts I had who had been squatting in abandoned basements for years also had been moved along. The city's most vulnerable had been cleared out of the city center by the very social forces of gentrification that my previous book had charted.

The clearing out of squatters from Bucharest's basements, tunnels, and sewer canals in the mid-2010s was accompanied by a steady buzz about the beautification and downward expansion of the capital that was familiar to those who had paid attention to Bucharest's earlier rounds of gentrification: columnists, real estate developers, and municipal officials talked about creating new Metro lines and updating old Metro stations, creating a new nightlife district in Old Town's cellars, and developing low-cost basement apartments and office rentals in the city's central neighborhoods, all of which were intended for the city's new middle classes. What I did not expect to find, and what cut against the familiar narrative arc of gentrification stories, was that this steady buzz was also accompanied by editorials, mayoral addresses, and city planning documents worrying over the presence of the city's growing middle classes in the city center. These voices insisted that the middle class's appetite for new cars and apartments had congested roads and crowded city blocks, polluted the air, and degraded the quality and character of public spaces. The new middle classes had not only physically overwhelmed the city to a breaking point, officials also voiced concern that their presence had become an eyesore that threatened the city's ability to attract future foreign investment. The proposed solution to these problems was simple enough: move the middle classes underground. For the next ten years, I followed the path this massive intervention took, alongside all of its insights and impositions, with the aim of understanding the underground: its material depths, to be sure, but also the social life that such depth makes possible.

As the work got under way, it became clear that I was not documenting a process of gentrification *for* the benefit of the middle classes, as gentrification efforts are readily assumed to do. Instead, I was tracking the gentrification *of* the middle classes. I found that Bucharest's new middle classes were getting swept up by the same structural forces of gentrification that usually target the poor in ways that were both familiar and novel. At its most basic, gentrification had cleared the most vulnerable *out* of the city center, but it was also working to move the middle classes *beneath* the city center. While the move underground no doubt sustained middle-class dreams of ascendency, it also introduced the subtle indignities of being moved below the city and out of the way of those who remained on top. Set against the backdrop of my earlier work, my intent with this book became to ethnographically theorize the intensification of techniques of class separation that privilege the wealthiest at the expense of the rest.

To that end, the chapters that follow track the loosely coordinated efforts of city planners and private developers, bureaucrats and entrepreneurs, architects and designers, as well as multi-national corporations to gentrify and expand the underground for the middle classes. The Introduction historically contextualizes Bucharest's turn toward the underground. It charts the political and economic shifts that brought about new middle classes in Bucharest, groups whose sudden ascent pushed the city's infrastructure to a breaking point. The opening chapter, "Lights and Tile," turns toward efforts to aesthetically repurpose the city's existing subterranean Metro stations, basements, and cellars in order to generate an atmosphere underground that would resonate with, rather than cut against, the tastes and preferences of the middle classes. Subsequent chapters, including "Station Kiosks," "Basement Apartments," and "Night Clubs," describe how the recycled underground facilitated middle-class dreams of working, living, and playing in an otherwise unaffordable city. The chapters also bring into view the many compromises and endless affronts that came with moving underground.

The fifth chapter, "Parking Garages," turns the book's focus toward the underground's expansion. It details municipal efforts to free the city's squares and boulevards from the middle class's illegal parking through the construction of subterranean garages. While planners approached the underground as an available repository that could absorb the material excesses that detracted from the city's charm, the simple fact is that the underground is not actually empty. It contains archaeological heritage that is legally protected. "Ruins" thus examines how private developers and government bureaucrats collaborate to empty the underground, producing the very subterranean repository that development demands. As the city extended ever further downward with ever greater optimism, "Foundations" points to how quickly such plans can become unstable. It discusses a series of widely publicized, catastrophic accidents that illustrate how the risks of turning toward the underground reverberate upward and outward across the city. The chapter also shows how the primary users of these spaces, the middle classes, are particularly vulnerable to these dangers.

The book's final three chapters consider the kind of public life that the underground makes possible. Chapter 8 focuses upon a hugely popular and civically minded series of advertisements inside the Metro called "The Digital Public Library." The eye-catching campaign purported to transform aging, milquetoast Metro passageways into visually stunning "libraries" where passengers could digitally download reading materials for their commutes. The chapter uses the Digital Library's holdings and limitations in order to bring into focus how the middle class was being cultivated underground. "Bomb Shelters" plunges to the very bottom of the city. Down there, at the lowest inhabitable level of Bucharest's major buildings, exists a system of government-mandated civil protection shelters. "Bomb Shelters" details both the inadequate quantity and the uneven quality of these shelters in order to chart out the limits of middle-class belonging to the city. The book ends in the city's municipal cemeteries, where the dream of escaping the subtle indignities of a middle-class existence lead the upwardly mobile to pursue prestige burial plots. Read together, the chapters of *Underground* bring into focus a new vertical ordering of the twenty-first-century city, one that supports dreams of ascendency for a rapidly growing middle class. These dreams, ultimately, are invariably colored by the degradations of being incorporated from below. The pages that follow are an invitation to reimagine the ordering of the twenty-first-century city from its foundations upward.

Underground

Introduction

This book gets to the bottom of contemporary urbanism. Metro stations, passageways and tunnels, basements and cellars, vaults and bunkers: the whole spectrum of urban life is found underground. Down there, people are living, working, shopping, riding trains, eating and drinking, studying, dancing, having sex, avoiding the rain, heat, and cold, reading, praying, chatting, cleaning, cooking, policing, taking pictures, surfing the web, listening to music, mining, boxing, meeting friends, getting high, steam bathing, undergoing teeth whitening, exploring, and of course decomposing; cemeteries are their own kind of city beneath the city. As the demands of global capital accumulation drive the cost of real estate to a premium, the kind of person going underground has changed from down and out to middle class. Additionally, the frequency of their trips underground has increased and the range of what they do once down there has expanded. To this end, a range of cities around the world have invested incredible sums of money to excavate, refurbish, and expand their undergrounds for the sake of their futures. *Underground* shows how these efforts have not only introduced new strata but also new aesthetics of inequality that are fundamentally shaping where and how the middle classes fit in the city.

City planners have long imagined that the cities of the future would grow downward. In 1910, for example, the Paris planner Eugène Hénard called for the development of underground roads and railways beneath the city's street level to "serve as a conduit for all the pipe systems, the removal of house refuse, and the transport of heavy materials and goods."[1] Writing from Philadelphia just a few years later, George Webster not only agreed with Hénard but also pushed the underground's potential further: "In the growth of a modern city, the number and character of underground structures are constantly increasing and changing and it is probable that the future will see the present ones supplemented by others now scarcely thought of." The need for a municipal body with absolute power to regulate subsurface structures seemed obvious enough to Webster; and yet, he added, "little thought has been given to the urban underground."[2]

But the advent of airplanes and skyscrapers quickly fixed the attention of these same planners and state regulators upward. "The maxim, *cujus est solum, efus est usque ad coelum* [whoever's is the soil, it is theirs all the way to Heaven and all the way to Hell] has had a long and honorable history," Laird Bell wrote in the *Illinois Law Review* in 1928.[3] But in an era of prolific upward construction, Bell foresaw the end of this paradigm of property rights. While grounded in legal precedent, the maximum's coordinates—an earth fixed to a heaven above and a hell below—implied a flat rather than round earth.[4] Moreover, modern advances in construction and travel had, in Bell's terms, thrown "the nature and extent of the rights of the owners of the soil projected upward" into question.[5] The economic potential of upward growth, particularly above the railroad tracks cutting through the valuable downtowns of the likes of Chicago and New York, appeared clear enough to developers.[6] Legal instruments for discerning "air rights"—and speculative markets on which to buy and sell them—quickly emerged to regulate the upward growth of these and other bustling metropolises worldwide.[7]

With imaginations gripped by the aesthetic wonders of the city's towering skyline, its downward development went startlingly unregulated, straight through to the present.[8] Master plans that meticulously coordinate the use of space from the city's surface upward do not incorporate the urban underground.[9] As one group of London-based architects noted, "We analyzed the [U.K.] planning laws and realized that they cover everything about the surface of the ground, but nothing beneath it. There was nothing whatsoever that could stop us from drilling all the way down to the south pole."[10] As land developers eager to maximize the value of their projects mine ever deeper underground, plans that imagined the underground as an empty repository ready to absorb the needs of development confront shifting geologies, networks of incomprehensively mapped utility lines and sewer canals, as well as the remnants of walls, foundations, passageways, and other legally protected heritage of the forgotten past. Once completed, these uncoordinated underground installations are especially costly to remove, creating obstructions that complicate any future efforts to systematically plan the city's immediate subsurface.[11] The obvious workaround, of course, is to drive future urban developments downward even deeper, beneath the detritus of the shallower layers.

Theorists have tended to give no more thought to the city's subterranean extensions than did city planners. While Georg Simmel brought attention to

the bridges and doors that organize the city's horizontal development by se-
lectively connecting city centers with their distant peripheries, he also
warned that the vertical relationships between buildings and their foundations are out of sight and therefore fall all too easily out of mind.[12] To be sure,
a robust literature about "the underground" did emerge in the mid-twentieth
century, but that literature employed the underground as a metaphor to describe the illicit economies and the subversive politics that animate the city's
most marginal and peripheral neighborhoods.[13] By extension, the people who
inhabited those marginal spaces—the so-called underclass—were taken to
be either the city's most vulnerable or its most improvisational and resource-
ful.[14] This flattening of the underground into a metaphor makes sense given
that scholars have been largely trained to see cities horizontally, through the
storied coordinates of center and periphery, city and country, local and global.
With eyes fixed on the horizon, spatially minded scholars have detailed the
dynamics of inclusion and exclusion effected through the production of
bridges and doors, walls, gates, and roads, for example, in ways that mark
some as *inside* and others as *outside* of the city.[15]

While revealing the brutal politics of exclusion from centers of privilege,
the metaphoric use of the underground is very much out of step with how
cities have actually extended underground, as well as for whom they have
done so. To put it simply, municipal and market actors in some of the world's
most dynamic and globally oriented cities are not investing billions to develop
and expand the urban underground to stash the down and out.[16] Instead—as
others have detailed so thoroughly, at times invoking the underground as
metaphor—processes of gentrification have pushed the very poor entirely out
of centers of wealth and privilege in order to make room for the middle
classes.[17] With their critical focus trained on the work of walls and gates in
excluding the very poor, scholars have not paid much attention to the way
floors and ceilings divide those within the city. As the density of urban development drives gentrification downward, renovating and expanding one
segment of the metropolis stacked beneath another, this book rethinks the
urban underground in order to account for the novel forms of vertical
fragmentation shaping where and how different kinds of people belong in
the city.

For the purposes of this book, then, the urban underground is not a
metaphor. First and foremost, it refers to the city taking shape beneath city
sidewalks.[18] The urban underground is made up of material spaces—such as
metro stations, basements, cellars, and passageways—that tens of millions

of people the world over pass through and rely upon every day.[19] The underground, from this perspective, is a subterranean extension of the city as a whole, one that is inextricably linked to the cityscapes above it. Importantly, though, this book also takes the relationship between the cityscape above and the underground below to be more than a geometrical fact. *Above* and *below* are also cultural coordinates of rank within mental or symbolic orders.[20] Yet, as this book shows, these orders are far from stable. The symbolic luggage of the underground is versatile, and it can be made to serve numerous different sets of values. Traditional cosmologies commonly extend positive connotations to what exists *above* while associating the grotesque and the profane with what lies *below*.[21] Unsurprisingly, the underground has traditionally been framed as a repository for all that has been cast down from the city above, including human waste, acerbic characters, forbidden pleasures, and poverty.[22] While enduring, these ranked coordinates have been softened by an alternative set of associations produced through modern planning and engineering. Techno-science has enabled a rethinking of the urban underground as adjacent rather than abject.[23] Sleek subterranean transit stations—rather than sewers and crypts—have become emblematic of the modern underground. The basement apartments that were once decried by nineteenth-century social reformers as "breeding places of disease" inhabited by only the most vulnerable have been repackaged and sold to striving middle-class professionals as garden-level apartments.[24] One consequence of this transformation is that the underground has become integral to the life and economy of the city rather than marginal to it.[25]

This book delves ethnographically into the lived space of the urban underground in order to make the claim that the underground produces and enforces a ranked ordering of urban life that is distributed vertically. In a moment of spiraling global inequality that has cleared outward signs of poverty away from densely developed centers of wealth and privilege, the novel perspective of the underground brings into focus how upstairs-downstairs dynamics increasingly characterize the experience of those who remain within the city.[26] As a nascent literature on the city's verticality has already observed, global elites do not just live in the city: they increasingly live *on top* of it.[27] Luxury residential and office tower developments have enabled the truly elite to retreat from the sidewalk into the security, privacy, and comfort of soaring city skylines. This trend certainly holds true in Bucharest.[28] This literature, however, has kept its gaze fixed on the penthouse, even as the same dynamics of property development that have placed elites *on top* of the city have also placed the middle classes in service to capital *on bottom*.[29]

Underground brings attention to the dramatic but overlooked developments taking shape in the city's depths. Down there, below ground, a new kind of middle-class public is being assembled. From the perspective of these aspiring, socially mobile citizens, the expanded underground is often a site of dreams and aspiration—hopes that underground developments might create opportunities for upward social mobility, allowing them to secure their place in the middle class. From basement apartments and offices to underground bars and clubs, and from small businesses in subway stations to grave plots in cemeteries, the urban underground makes accessible a relative sense of prosperity and prestige. Underground, the growing middle classes can find the space to move through, live, work, and play in already overcrowded city centers. Yet these opportunities are not without costs. Underground spaces are heavily trafficked, semi-public sites designed to make the labor and consumption of the middle classes available to businesses within the city center without their growing presence physically overwhelming the city center. And, at the edges of these efforts to make the underground an aesthetically appealing place for the middle classes to go, the dark, dank traces of the underground's muck and mire always threaten to seep through, presenting the subtle indignities and sense of slipping status that come with being incorporated into the polity by way of the basement door. It is at the intersection of these two experiences of the gentrified urban underground—dreams and degradation, aspirations and anxiety—that a new middle class is being formed. Rather than a site of radical or marginal public culture, then, the pages that follow show how the underground is foundational to the formation of a global middle class.

Bucharest

To bring the emergent urban underground into focus, this book turns its attention to one of Europe's most densely developed capital cities: Bucharest, Romania. In the last few decades, Bucharest has gentrified and dramatically expanded its underground in response to a changing economy, growing middle classes, and an increased appetite for foreign investment. Although a peripheral city in the urban studies literature, Bucharest's expanded underground renders visible, without disguise or subtlety, new patterns of separation in cities—patterns that place different classes of people beneath and on top of each other. The relative success or failure of these efforts at coordinating competing claims to a busy downtown, this book shows, hinges upon the ability of developers to aesthetically repurpose the underground to resonate with the dreams of upward mobility such that the indignities of being underground recede.

To comprehend how these dreams of upward mobility emerged, a working understanding of Bucharest's changing politics and economics, as well as of the rapid growth of the city's new middle classes, is necessary. For much of the twentieth century, Bucharest developed as a center of socialist industry. The Romanian Communist Party (Partidul Comunist Român, PCR) came to power in 1947 and instituted a program of "systematization" (*sistematizare*), which coordinated territorial planning and economic development.[30] In the name of establishing the proletariat that the Communist Party's politics demanded, Romania's longtime dictator, Nicolae Ceaușescu, actively pruned the countryside while investing in cities. A program of "village consolidation" forced waves of rural peasants to migrate to the capital, doubling Bucharest's population as they satisfied the labor demand of its new factories.[31] Planners established new residential zones, such as Drumul Taberei, Berceni, and Titan (Balta Albă) to accommodate rural to urban migrants.[32] Urbanists in this period encouraged population density by growing the average construction height of the city to six stories.[33] While the capital's established and connected residents congregated in the city's center-north districts, city planners relied upon public transit to promote equality of access between the new residential zones and the capital's cultural, civic, and medical resources.[34] While industrialization generally improved the quality of life in Bucharest in the 1960s and 1970s, the Romanian Communist Party's mismanagement of the economy eventually led to widespread austerity during the 1980s.[35] During the 1980s, brownouts and breadlines became regular features of urban life, causing frustrations to boil over. On Christmas Day 1989, an anti-Ceaușescu uprising culminated with the execution of the longtime dictator and his wife, bringing an abrupt end to Romanian socialism.

Western reformers promised a quick transition out of socialist-era austerity and into the prosperity of the global economy.[36] To that end, the Romanian state sold roughly 4.3 million housing units, as well as majority shareholder status in all of its non-essential industries, to private individuals.[37] Officials also opened Romania's borders, allowing its citizens to move abroad while inviting the likes of Coca Cola, McDonald's, and Nike to move in. However, rather than streamlining operations and accelerating their output, the privatization of factories caused them to stall almost entirely. By 1994, around one million workers—a quarter of the industrial workforce—had been abruptly forced out of the labor market, Romania's already low real incomes had dropped by more than 40 percent, and the value of pensions plummeted.[38] By 1999, the share of the population living below the national poverty line had doubled to over 41 percent, and male mortality rates had increased across the region.[39] Bucharest's potholed streets and crumbling facades emerged as concrete signs to international observers of Romania's financial hardship and fiscal neglect. From the French daily *Libération* to the *Economist* and the *New York Times*, editorialists depicted Bucharest as a city overrun with feral dogs, illicit street vendors, and the homeless.[40] One *Newsweek* columnist cited Bucharest's poverty and dilapidation as hard evidence of Romania's failed transition in order to raise concerns about the country's fiscal mismanagement, political corruption, and, ultimately, its fitness for EU accession.[41]

Those who could leave Romania's collapsing economy did. The country's population fell from 22.8 million in 2000 to 19.6 million in 2017, as more than 20 percent of its working-age population emigrated abroad.[42] While recording one of the European Union's highest shares of emigrants, Romania only attracted about 100,000 migrants, mainly workers from Nepal, Bangladesh, Sri Lanka, and Turkey.[43] Not surprisingly, remittances emerged as an important element of the Romanian economy. Between 2004 and 2016, Romania received an average of 3.5 billion euros in remittances, representing 2.8 percent of its GDP.[44]

Prospects began to dramatically change in Bucharest, however, at the turn of the millennium, with Romania's accession into the European Union (EU) imminent. Foreign investors who had written off Romania's socialist-era industry saw potential in making Bucharest a back-office hub. Romanians' fluency in languages, science, and technology as well as the capital's comparatively low cost of living and proximity to Western Europe prompted the likes of Microsoft, IBM, and Oracle to near-shore their back offices to Bucharest rather than offshoring them to India.[45] Foreign investment surged tenfold between 2000 and 2010, setting off a construction boom in Bucharest's center-north business district.[46] While Romania's population plummeted due to emigration, Bucharest continued to grow.[47] The suburbanization of neighboring villages further contributed to Bucharest's density as suburbanites commuted into the city for work.[48] The number of new buildings in Bucharest nearly doubled between 2002 and 2006, and in the decade following the 2008 global financial crisis Bucharest's stock of modern office space came close to doubling again.[49] High-rise buildings mushroomed in such close proximity to one another that newer buildings prevented existing ones from receiving direct sunlight.[50] Amidst the tidal wave of foreign investment into Bucharest, business services ballooned to account for 70 percent of the city's economy.[51] Romania quickly transformed from one of Europe's poorest countries into one of its fastest growing economies.[52]

With this new economy came new middle classes—or at least talk of them. Yet, though such talk circulated widely in Bucharest, the precise socioeconomic coordinates of the group being identified remained difficult to discern. This makes sense given that the term *middle class* has been alternatively described within the social sciences as both "fuzzy" and "breathtakingly vague."[53] Simply put, middle class is used too loosely in Bucharest, and elsewhere, for it to function as a coherent, sociological category. For all its admitted conceptual messiness, however, middle class is very much a collective subjectivity in circulation in Bucharest and beyond, with the presence and force necessary to dramatically shape urban life. For example, government programs purport to invest in the middle class, city planners and private developers execute projects for the middle class, designers stage these spaces to appeal to middle-class audiences, and city residents locate themselves and others within (or outside of) the middle class. These everyday practices bring the middle classes into being as a durable social reality or "public," as scholars have put it.[54]

This book contends that the urban underground played a central role in defining what it meant to be a part of the middle classes in twenty-first-century Bucharest. As an emerging yet ill-defined group, Bucharest's growing assemblage of upwardly mobile property owners, office workers, educated urbanites, top public employees, and university students attracted much of the attention of market and municipal actors alike, groups who saw the former as valuable for their labor and consumption.[55] At the same time, however, the upwardly mobiles' growing numbers and rising consumerism threatened to physically overwhelm a city that had already been pressed to its infrastructural limits. As the chapters that follow demonstrate, city planners and private developers attempted to negotiate this dilemma by attracting the city's young professionals underground—by building and marketing basement apartments, by opening new subterranean nightclubs, fast food restaurants, and coffee shops, by making improvements to subway stations and parking garages. These projects relied upon a set of commodity aesthetics that did not attract so much as actively compose the very middle-class public that foreign investors demanded. While glossy finishes and developer brochures framed these underground developments in an appealing way, their budgets were never unlimited, the effort never fully exhaustive. Gentrification rarely overcame the material conditions associated with life underground. Musty smells, flickering halogen lights, and grimy floors marked the limits of investment in the city's subterranean extensions as much as in the new middle classes. The underground's discomforts and subtle affronts cast their shadow upon middle-class fantasies of ascendency such that a degraded sense of belowness threatened to take hold. Bucharest's emerging group of back-office workers, ultimately, became a coherent and marketable middle-class public through the dreams and degradations evoked by the processes of vertical, aesthetic ordering that this book describes.[56]

The formation of Bucharest's new middle classes occurred against the backdrop of staggering inequality both within Romania and across the EU. In 2016, when roughly a quarter of Romanians earned less than €2 per hour, the steady inflow of foreign investment raised the average net income in Bucharest to €7,884 per year, nearly twice that found in Romania's poorest counties.[57] The divide between the median hourly earnings of high- and low-skilled workers in Romania grew to the widest in the EU.[58] As residents of the capital spoke confidently about their quality of life pulling away from that of the rest of the country, they were also very much aware that the wages they earned in the Bucharest back office lagged far behind those in the front offices of Western Europe. Median gross annual earnings at the same time were €30,202 in Belgium, €24,850 in Germany, and €22,790 in France.[59] Moreover, Romanians earned their comparatively lower wages by working longer hours than anywhere else in Europe.[60]

These new middle classes used their rising wages to buy into newly constructed residential developments along Bucharest's former industrial periphery, in neighborhoods such as Militari, Pantelimon, and Berceni.[61] They still, however, commuted to and worked and consumed in the city's centers of business and commerce. To move between home, work, and the night out on the town, Bucharest's new middle classes turned en masse away from public transit and toward personal automobility. Spiking car ownership rates sent 1.2 million cars circulating through a socialist-era roadway system designed for 180,000, bottlenecking roadways and immobilizing traffic.[62]

As construction and car ownership boomed, a very particular kind of mobility problem played out across the city, one in which the materiality of middle-class consumerism ground against Bucharest's socialist-era infrastructure in ways that threatened the city's continued ascendency. Evidence of the growing mobility problem abounds: during the course of this research, the Dutch navigation company TomTom named Bucharest the most congested city in Europe and the fifth most congested city in the world, with daily commutes worse than the storied motorways of Los Angeles, Beijing, and Istanbul.[63] Bucharest's interminable traffic left the city with the poorest air quality in the EU.[64] When not in motion, those who navigated Bucharest's socialist-era roadway system were left with nowhere to park, leading to a staggering one million illegally parked cars overrunning the city's sidewalks, squares, and parks.[65] Not surprisingly, as illegal parking pressed pedestrians into overcrowded roadways, Romania's rate of pedestrian fatalities shot to the very top within the EU.[66] And dire as these trends were, planners predicted that Bucharest's mobility problem would only get worse.[67] Officials worried openly about Bucharest's growing "dysfunction" hindering its ability to continue to attract the very foreign investment that had made middle-class growth possible in the first place.[68]

Amidst interrelated concerns about mobility, aesthetics, and development in the city, the mayor of Bucharest proposed to remake the heart of the capital by gentrifying and expanding its underground for the middle classes.[69] Touted as the biggest urban development project in the history of Romania, the plan proposed an €8.5 billion extension to the Bucharest Metro system that would double the number of stations, triple the length of track, and connect the airport to the city center.[70] In addition to expanding the Metro system, the plan called for the gentrification of the Metro's existing stations. As happened in other post-socialist contexts, a pronounced presence of unhoused persons had taken shape underground during the 1990s and 2000s, which led Bucharest's upwardly mobile to associate the Metro system with low social standing.[71] Heightened policing soon cleared so-called "street children" and panhandlers from the Metro system.[72] A subsequent police action also brutally evicted a longstanding squatter encampment that had taken shape in the subterranean heating canals adjacent to the Gara de Nord railway and Metro stations.[73] These clearings and evictions created the possibility for the underground's aesthetic repurposing for the middle classes that this book follows. At the same time, the municipality earmarked an additional €311 million to build a new constellation of underground parking garages in central Bucharest.[74] Private developers followed suit, investing millions to convert the city center's basements and cellars into grocery stores, bars, restaurants, apartments, and offices.[75] The newfound value of the underground quickly spread from the metropolis to the necropolis, with the value of crypts in the city's oldest cemeteries reaching parity with studio apartments (*garsonierā*) in the city center.[76]

By gentrifying and expanding Bucharest's underground, planners hoped to ease demands on the city above, freeing central Bucharest to transform into what it has long aspired to be: "The Paris of the East."[77] Becoming more like Paris, Bucharest's planning documents argued, was imperative to remaining attractive to foreign investors.[78] Such a project hinged as much on generating a Parisian atmosphere as on improving the city's technical and logistical capacities. "The heterogeneous and eclectic character of the center of Bucharest must be emphasized in order to create an urban identity that the inhabitants are proud of and which will attract tourists and investors alike," stated the Integrated Development Plan for Central Bucharest.[79] With Paris as the diagram, planners looked to clear Bucharest of the traffic, smog, and illegally parked cars as well as to showcase the architecture of its city center and stage its ground floors with just the right storefronts. Through evoking Parisian dreams, planners hoped that foreign entities would continue to invest in Bucharest as an "untapped European metropolis," one where streets were not only walkable but lined with quaint cafés, bars, and restaurants.[80] To affect what might best be described as its Parisian "edifice complex," city planners and bureaucrats sought to create an underground that could absorb the excess of people, things, and activities that didn't quite fit the city's intended image.[81]

Such a monumental undertaking did not start from scratch. Since its medieval origins, Bucharest's underground has always been an integral stratum of urban life. The city's earliest settlers, for example, built a constellation of arched cellars supported by one-meter-thick walls beneath historic Bucharest.[82] Travelling merchants relied upon the fortified subterranean galleries of the Old Princely Court (Curtea Veche), which was the fifteenth-century palatial home of Vlad the Impaler, as well as those of nearby mansions and inns, to secure their trade in valuable jewels and coins.[83] The pubs that entertained these traveling merchants as well as the homes of the artisans who did business with them, meanwhile, used similarly cavernous cellars to store wine and brandy (*țuică*) from nearby vineyards and orchards.[84] A labyrinth of subterranean tunnels also crisscrosses the heart of old Bucharest. They emanate out of the basements and cellars of historic Bucharest and were used to escape invading Ottomans.[85] Built behind sliding bookcases and ornately carved desks, for example, some of these escape tunnels emptied into stately galleries described as having the feel of an underground castle.[86] In the mid-nineteenth century, civic leaders established the Șerban-Vodă (Bellu) Cemetery, Bucharest's first municipal cemetery, to modernize and order burials.[87] Bellu cemetery became the resting place of Romania's pantheon of political and cultural figures, who were commemorated with dramatic crypts adorned with ornate headstones that extended aboveground and would, a century later, cast the cemetery as an open-air museum of modern art.[88]

In the twentieth century, the role of the underground expanded further still. In 1935, the Romanian Dadaist Marcel Iancu proposed a new division of space for Bucharest, one that placed residents in apartment blocks extending high above the city surface while relocating transportation beneath it.[89] The vision of a garden city with residents moving about underground, however, did not crystalize until the socialist era. With street trolleys and trams crowded to the breaking point, city officials opened the Bucharest Metro in 1979.[90] Romanian planners noted with pride the immense volume and scale of its central underground stations, which were intended to bring the feel of a main square underground.[91] Trafficking in the language of modernity and efficiency, planners framed the Metro as a new image of the future.[92] While the government built toward a socialist utopia with the construction of Metro stations and tracks, it also prepared for that dream's unraveling through a parallel tunneling effort: high-ranking officials in Bucharest constructed and maintained secret underground tunnels to secure their own egress from the capital in the event of a Soviet invasion or a populist uprising.[93] Following the demise of the Ceaușescu regime, accounts emerged of an underground labyrinth stocked with armaments, food, and motorcycles as well as a barracks and an underground office bunker.[94]

In the twenty-first century, the underground has become ever more important to the life and prosperity of the city. Its expansion has been foundational to producing an aesthetic ordering of the city, one intended to appeal to foreign investment by constituting an available pool of middle-class office workers and consumers without stepping upon the city's Parisian charm. By holding out the possibility of a share of prosperity that reflects the city's broader development, the aesthetically repurposed underground opens up as a site of middle-class aspirations—even as the inevitable limitations of subterranean life threaten disillusionment. Rather than being the site of abjection and abandonment, as metaphors of the underground suggest, the gentrified and expanded underground might best be theorized as strategically productive.[95] And the need to theorize this emergent underground is apparent. Around the world, cities are turning to the underground to negotiate the challenges of hyper urbanization, climate change, and environmental risk.[96] The pace of urbanization in major American and European cities is driving expansions of existing Metro systems, while cities across Asia are rapidly introducing new underground transit.[97] From Helsinki to Singapore, municipalities are beginning to incorporate the underground into their masterplans.[98] In glossy reports, consulting firms boast of the development of underground farming in Paris, the Lowline park in New York, nightclubs, museums, and movie theaters in London, public swimming pools and hockey rinks in Helsinki, and a seven-layer shopping center beneath Mexico City.[99] In Hong Kong, development projects rise above and dip below the city surface with such regularity that the sidewalk has ceased to be a meaningful reference point.[100] The scale, scope, and ambitions of such projects have led to splashy headlines in venues such as *Smithsonian* magazine declaring the underground "the next frontier in urban design."[101] As the pursuit of global capital accumulation drives the development of cities not just outward toward evermore distant peripheries and upward into ever higher skylines, but also downward beneath the sidewalk, this book considers the dreams and degradations of Bucharest's gentrified underground to provide needed ethnographic insight into the potentialities and pitfalls of the twenty-first-century city's heightened verticality.

While the main text lends what is intended to be a broadly accessible narrative voice to the argument, substantive notes located at the back of the text firmly stabilize the narrative in evidence and theory. Advanced scholars interested in getting to the bottom of the book's finer nuances should tend to the groundwork below as much to the narrative above. In this way, *Underground* reflects something of the city that it describes.

CHAPTER 1

Lights and Tile

Amidst growing concerns that unchecked development was pushing Bucharest to a breaking point, city hall turned toward the underground to restore the city to working order.[1] With Paris and Brussels as its guide, in 2008, the mayor proposed a dramatic expansion of the Bucharest Metro system. Billions were poured into excavating the city surface so as to create an underground metropolitan transport network that would ease congestion by connecting the city's center with its peri-urban reaches.[2] While the underground's physical expansion could be engineered easily enough, planners noted with concern that city residents would still have to be convinced to go underground. Remaking the city from its foundations upward was therefore as much a matter of adjusting the underground's aesthetics as of expanding its functional capacity. And city planners noted that there was considerable work to be done. "Stations are poorly illuminated, finished with dark and uninviting materials," they freely conceded, "and while these factors do not affect the efficiency of the system, they do decrease [the Metro's] attractiveness, especially for passengers who have the choice between public and private transit."[3] Such public angst over the Bucharest Metro's socialist-era character unfolded alongside discussions and grand initiatives in cities such as Paris, London, and New York about the developmental imperative to improve the user experience of their own aging undergrounds.[4]

Beyond matters of lighting and finish, administrators in Bucharest also voiced concern that the commercial amenities found underground were vestiges of the penny capitalism of the initial years of Bucharest's post-socialist transition and, therefore, would potentially offend emerging more up-market sensibilities. The ubiquitous presence in Metro stations of convenience kiosks facilitating grab-and-go purchases of prepackaged snacks, deodorant, and

Figure 1. Aged Metro, 2017. Photo by author.

racy lingerie emerged as the most prominent case in point (see Chapter 2). As one senior administrator for the Metro bluntly stated in 2014, Metro stations needed to "get rid of the 'gypsies,' get rid of the intimate lingerie vendors, and get rid of the pastry shops filling stations with suspicious smells. [In their place, administrators] would like to see modern stores, and eventually multinational brands."[5] As the language of "misery and chaos" (*mizerie si haos*) came to characterize Metro stations in the popular press, Bucharest's upwardly mobile increasingly associated the underground with low social standing, as has been the case in other post-socialist cities, from Moscow to Tbilisi.[6] Conversations with workers in Bucharest's back offices confirmed as much. Convenient as the Metro may be, young office professionals explained to me, they would rather sit in traffic in the comfort of their own cars than pass through "dirty" Metro stations and ride in "overcrowded" trains.

To make the underground into a compelling place for the middle classes to go, municipal and market actors worked to refine its "atmosphere," which Gernot Böhme insightfully elaborates as "the emotional tinge of a space."[7] By acting upon the underground's material qualities and design—by tweaking its lighting, adjusting its acoustics, dissipating its humidity, and then improving its finishes—city planners and private developers worked to recast the atmosphere of the underground in a more appealing light. Through a sustained aesthetic intervention into the underground's staged materiality, these actors attempted to change the character of the underground's Metro stations, basements, and cellars so as to more positively resonate with the elevated lifestyles of the upwardly mobile.[8]

This chapter examines the effort of entrepreneurs, designers, and planners to attract the middle classes underground through the staging of atmosphere. It examines the kinds of aesthetic interventions made to rework the city's dingy Metro stations, basements, and cellars so that the underground would no longer trigger status anxieties among the upwardly mobile but would instead act upon dreams of ascendency into the global middle classes.[9] In what follows, I examine three contrasting strategies for enticing the middle classes to move underground. The stakes of staging the underground's materiality in alignment with middle-class sensibilities were high: nothing less than the successful reworking of Bucharest's class structure hinged upon the creation of a suitable atmosphere.

Convenience

For two decades, the Romanian franchise of McDonald's operated two restaurants in the capital's most central zone, Union (Unirii).[10] The first opened onto Union Square (Piața Unirii) at street level, while the second was located directly underground inside the Union Metro Station (Statia Unirii). The two McDonald's were almost perfectly aligned, with one restaurant located more or less exactly above the other. But they were so different in character that a customer could believe, going from the restaurant above to the other below, that they were entering another world.

The restaurant above is located along the square, on the ground floor of the Union (Unirea) Shopping Center. Passing through the restaurant's double doors, one crosses into a heavily air-conditioned entryway that opens into a large atrium, which is then sub-divided into more intimate seating areas. Tabletops at this restaurant are wood-stained and their surrounding benches, chairs, and stools are softly padded in brown or cream faux leather. Dark, wood-stained walls, decorated with framed prints, encase seating clusters. Upbeat pop music plays in the background—softly, so as not to intrude on conversation—and a bank of registers sits opposite the entrance. As Romania's flagship McDonald's restaurant, customers order from across the franchise's

Figure 2. Union Square McDonald's, 2015. Photo by author.

Figure 3. Union Station McDonald's, 2016. Photo by author.

comprehensive menu. A McCafé station is located off to the side, where baristas in fitted uniforms serve espresso-based drinks in ceramic mugs and designer desserts on warmed plates. This McDonald's is a beacon of excellence, on par in the quality of its customer experience with Starbucks or Caffè Nero. The atmosphere invites customers to slow down and settle in, offering a respite from the hustle and bustle just outside its doors.

The other McDonald's was found, until 2019, in the vestibule of the Metro station below. When entering from the station's comparative darkness, customers squinted, their eyes not yet adjusted to the bright halogen lights, the glare from which reverberated off of rigorously cleaned, glossy white laminate tables and countertops. Thumping dance music drowned out all but the loudest sounds of the Metro's coming and going trains. Metallic railings, subtly disguised as plant stands, corralled customers toward a bank of registers, where customers ordered from a pared down menu of core offerings. While the majority of customers took their orders to go, seating areas off to the side accommodated those in need of a place to eat their quick bite. Straight-backed, unpadded chairs discouraged settling in. The volume of music impeded conversation. Luminously clean and functionally efficient, the atmosphere in this McDonald's encouraged customers to stay in gear and to get back on the go. And yet, what at first glance might appear as two different worlds proved on closer examination to be strategic variations of a single McDonald's experience intended to cater to different classes of people.

"Do you see the kinds of tables and chairs over there?" asked Cristian, a regional facilities manager for Romania's McDonald's franchise, as he gave me the tour of the Union Station restaurant in the summer of 2015. "Their number and arrangement, the lighting, this music," he continued, "none of these are fixed. Everything that you see is a response to the observed needs of clients at this location." Management—from the corporate office down to the individual restaurant level—was unwavering on this point: the palpable difference in character between McDonald's restaurants was neither technical nor incidental but rather deliberate and strategic. As Alexandru, a senior manager for the Romanian franchise of McDonald's, maintained, "The central premise of our business is to pay close attention to the customer segment of each restaurant and then adapt what we do to meet the specific needs of that restaurant."

Through its attention to aesthetics, management strategically acted upon the relationship between a restaurant's material conditions and the subjective states of its targeted customer base to generate a specifically tailored McDonald's experience for each location. To tailor the experience to the customer segment of each restaurant, management tweaked the materials, lighting, and sound of its floor space to strike just the right aesthetic, to generate just the right mood. It is a statement on operational practice that stands in stark contrast to the way McDonald's has been popularly theorized as a symbol of homogenization.[11] "With our customer experience," Alexandru continued, "we have never worked with a one-size-fits-all model. That would make no sense, because what customers in one location want or need is not the same as customers in another location." The comparatively functional, verging on austere, quality of the Union Station McDonald's, Cristian and other franchise managers insisted, was not an imposed constraint of operating under- as opposed to aboveground.[12] Cristian maintained that the McDonald's franchise was completely capable of recreating the same character and comfort installed in its Union Square restaurant down below in Union Station. "In both instances," he stressed, "we control the quality of the lighting, the air filtration, the sound. From the standpoint of producing the customer experience, being underground does not factor into design. It is not a consideration."[13] The difference in character across the two restaurants was instead, by management's account, a strategic response to the kind of customer that each restaurant served.

"The big difference between our two Union restaurants is the type of clients," explained Liviu, a regional operations manager for McDonald's. We sat at a table down in the Union Station restaurant. "On the square," he continued, pointing above his head, "there are people with money, people who come wanting to be entertained, to have fun shopping. They are businessmen and their families. But down here," Liviu went on, returning his hand to the table and leaning forward on his elbows so he could still be heard while speaking in a softer voice, "it's a different customer segment. The clients down here are trying to get to work, or school, or home. They're in a rush to get somewhere else. And they're using the Metro rather than their own car. They're middle class." Aboveground, things were different. "There is a wealthier class of client visiting Union Square," Cristian told me. "We know that it serves businessmen, bureaucrats, and lawyers from nearby offices as well as their families who spend time at the Union Mall [Unirea Shopping Center]. They come wanting to escape the office, the home, and they want to be entertained."

By creating these distinct experiences, McDonald's not only catered to but also created the class differences that existed between the Union Square and Station locations.[14] The experience at Union Square was staged for the higher-end professional who is afforded hour-long lunch breaks and has the discretion to schedule meetings outside of the office. The restaurant was designed to put these professionals at ease, and it was fabricated with a certain comfort and quality of amenities so as to attract this high-spending segment. Once inside, the restaurant's atmosphere then worked meticulously to prolong their visit so as to increase the likelihood of additional purchases. "We want these customers to benefit from all of the gadgets in the restaurant," Liviu explained, gesturing to a row of iPads available for web surfing, "to sip an espresso, and then go back and order a special dessert." To keep the square's wealthy clientele in house longer, the McDonald's experience on the square was composed of seating options that were not just expansive but cushy, the lighting warm and the soundscape soft. Management staged Union Square to encourage its upmarket clients to slow down and settle into comfortable seating in the hopes of triggering continued spending.[15] The opposite was true down below.

"The design in the Metro is L.I.M.—Less Is More—to the extreme," explained Alexandru at our follow-up meeting.[16] "Customers at Union Station are middle-income people in a hurry," he continued. Instead of attracting higher-paid professionals, the restaurant underground was designed for students and workers on the lower rungs of organizational charts. Rather than getting to exercise discretion over their arrival time like the managerial classes, this rung of clients were expected (by those same managers) to clock in regularly and promptly at their desks. "They need to eat but they don't have the time to sit around in a restaurant. And so, the restaurant experience in the Metro is all about speed."[17] From management's perspective, the staged materiality of its Union Station restaurant—with its barebones seating, energetic music, and bright halogen lighting—did not reflect a diminished sense of hospitality for the comfort of its down-market clients. On the contrary, the staged materiality of Union Station reflected corporate's intensive efforts at market segmentation and profiling. The "L.I.M." aesthetic of the station's restaurant sought to align the materiality underground with the life conditions of the lower rungs of the middle classes. Though it would present an unpleasant (if not impossible) option for the lawyer looking to meet with clients outside the office, the restaurant underground, by design, worked to get the harried worker bee fed and back out the door as quickly as possible.

The case of McDonald's makes clear that the dark and dirty underground can be imagined anew through focused attention to a site's staged materiality.

It demonstrates how the underground can be designed to cater to one segment of the middle classes, rather than as a repository for the poor. Notably, such a project does not entail the reproduction down below of the character and comforts of the world above. After all, the McDonald's franchise imagined different classes of people staying on top of the city as opposed to moving below it. This act of class separation opens up the opportunity to eat a Big Mac for lunch to the overworked and underpaid visiting the Union Station McDonald's, for example, just as the managerial classes do in the restaurant located just above the formers' heads. While the underground brings the pleasures of a Big Mac within reach, the experience is nevertheless coupled with the discomfort of scarfing down that sandwich on the go rather than getting the opportunity to savor the experience at one's own pace. More than catering to class differences between the city above and below, the shift in design strategy creates class differentiation.

Cool

The call to aesthetically repurpose the underground for the middle classes extended beyond the Metro system. Alongside municipal efforts to move the daily commute underground by generating a more appealing atmosphere inside Metro stations, developers and entrepreneurs worked to redesign Bucharest's basements and cellars to facilitate the growing middle classes' desires to live, work, and play in an already overcrowded city center. "The market for real estate underground has radically changed in the last ten years," explained Sorin, a senior executive for a Romanian development firm, in 2015. Sorin attributed the surging interest in converting the city's basements and cellars into cafés, bars, offices, and apartments to the development of a nightlife district in Bucharest's Old Town (Centrul Vechi) neighborhood. The area features a dense concentration of bars and restaurants where upwardly mobile university students dance late into the night alongside office workers and foreign tourists. The development was immediately lucrative, drawing fierce competition for space within Old Town's densely developed corridors. "In Bucharest," Sorin continued, "it's rare to see a sale price exceed €3,000 per square meter, but in Old Town, prime commercial spaces are going for €9,000 per square meter, which is just unheard of." With space at a premium, the most entrepreneurial moved to aesthetically repurpose the

Figures 4A and B. Converted cellar offices, 2014. (top) Preparing cellar conversion;
(bottom) completed cellar conversion. Photos by Palat Noblesse Group.
Photo reproduced with permission.

neighborhood's storage cellars. "The financial incentive is clear," Sorin explained, "by heading into the basement, you can pay a smaller rent. And then what's become apparent is that with the right design, being underground becomes a bit more attractive, a bit more interesting for clients." Designers drew upon the metaphor of the underground as a signifier of counterculture to strategically frame the underground's aesthetic. With the right intervention, the cellar's grit became the source of its claim to cool.

As they did with the imagined relationship between main squares and the Metro stations beneath them, designers approached basements and cellars for a middle-class clientele that they imagined to be distinct from those who stayed atop of the city. "It's the same story in Venice," Lorenzo, an Italian interior designer who trained and worked in Bucharest, explained to me in the winter of 2016. We sat in his studio in central Bucharest reviewing photographs of his past projects. "The restaurants that are easy to find are for tourists and the moneyed. The interiors are flashy, everything on the menu is expensive, but the experience isn't necessarily memorable," Lorenzo assessed dismissively. "The interesting experiences are actually found underground. There you find where the people actually living and working in the city go." In Lorenzo's analysis, the underground is not the cheaper approximation of life on the city square but rather its cooler alternative. The high-gloss, high-cost experience found above, from his perspective, is a kind of scam where those with an excess of economic capital happily pay high prices because they do not have the cultural capital to know any better. "The restaurants, bars, and cafés are more interesting underground because people have to make something beautiful out of the basement," Lorenzo continued, admiring images of his own efforts. Shifting from analysis to design philosophy, he added, "With my work, I see the city like a treasure map, with the excitement buried down below." While Lorenzo's perspective is obviously self-serving, his point is not without merit. Framed in just the right light, polished with just the right touch, nightlife has thrived in Old Town's basements and cellars (see Chapter 4).

Those tracking development in Old Town noted two trends pertaining to the middle classes. "One segment of clients is headed toward basement clubs—they're younger, not as wealthy but more pretentious than those staying on ground level," Sorin characterized. "The drinks are very cheap in underground clubs, the image is countercultural, and they can party as late as they want. The basement and cellar clubs have become their own appealing

Figure 5. Cramă, 2018. Photo by author.

thing." At the same time, Sorin noted, an older set of middle-class professionals and tourists have gravitated toward the newly renovated cellars (*beciuri*) of Old Town's medieval inns such as Hanu' lui Manuc and Crama Domneasca. These inns, like others built in the same period, were constructed as true fortresses, with one-meter-thick brick walls, shops on the ground floor, and rooms for travelers and traders on the upper floors.[18] In their cavernous cellars, copious amounts of wine and brandy were not only stored but also served late into the night to wealthy merchants.[19] Reimagined in recent years as restaurants, the cellars now serve as event spaces. "They're beautifully done and they can handle large groups: office parties, family reunions, tourist groups," Sorin explained. The sharp contrast Sorin drew between clubgoers and corporate office parties, with each demographic being differently situated underground, speaks to the diversity of the middle classes themselves. As with the case of McDonald's, market actors designed basements and cellars to create as much as to cater to different segments of the middle class. As the club and the office party moved underground, they both sat in opposition to Bucharest's high-earning professionals, who had turned upward, to the office and residential towers that placed them above it all.[20]

To generate an underground that appealed to the varying needs and desires of the middle classes, Lorenzo played upon its particularity. "The basement in Bucharest has a lot of history," he continued from his design studio. Indeed, spending time underground is far from unprecedented in Romania. In the eighteenth and nineteenth centuries, the cellar-level (*pivniță, beci*) of Bucharest's wealthy merchant villas was typically a living space utilized for food storage, cooking, and cleaning while the upper floors were largely reserved for entertaining.[21] These cellars can have complex brickwork and vaulted ceilings that give the spaces a sophisticated look, especially when unfinished. "People have made basements upon previous buildings, and renovated rooms out of other rooms. This leaves you with layers of history that change styles. And so, in the basement you can find something distinct to work with, to draw out—you can make 'other kinds of possibilities' than can be done with the portions above."[22] Whereas McDonald's management conceptualized the underground as a kind of blank geometric space that could be remade to one's specifications, Lorenzo found that Bucharest's basements and cellars opened up a new creative domain. Visually separated from the rooms above and the street beyond, the basement level allowed for the creation of spaces that reflected back on their own history. The cellars' walls and

ceilings evidenced the craftsmanship of earlier eras, and so, rather than eras-
ing or covering them, Lorenzo instead turned them into the focal points of
his contemporary interiors. "You have to approach basements and cellars as
completely different from the aboveground floors because you don't have the
same height, the same light, the same sound, the circadian rhythms or views,"
he continued, flipping through his catalog of bars and clubs, offices, spas, and
meditation rooms. Each represented its own little world set unto itself. "The
goal is to draw attention to the beauty in what is already down there and to
encourage uses that are enhanced by playing with light and shadows. It's
about designing toward pleasure and excitement."[23]

As the nighttime economy in Old Town demonstrated the commercial
viability of basements and cellars, an increasing number of owners of villas
beyond Old Town moved to cash in on their own subterranean spaces. "Unfortunately, most of the owners don't know how to exploit their basements,"
Toma, a real estate agent specializing in central Bucharest, told me in 2016
while we toured a luxury neighborhood called Dorobanți. "They are not properly investing in their design, and it's keeping renters from seeing the potential. I don't waste time with those," Toma said, affirming the imperative to
appropriately stage the materiality of basements and cellars to draw in the
middle classes. "But, if you invest the money to make it nice," he assured me,
"you will absolutely attract the educated, the creative, who prefer to be here
[in historic Bucharest] rather than priced out to Obor [a historically working-
class neighborhood]. This segment appreciates the architecture, the amenities, and they want to be surrounded by a certain quality of person." As we
walked, Toma explained that renovated basements had become particularly
attractive sites for small businesses. "People are less anxious about natural
light where they work as opposed to where they live," he psychologized. He
noted the proliferation of signs for medical, legal, and architectural firms af-
fixed to the basement windows of the neighborhood's historic villas. "These
are the kinds of businesses that don't bother the tenants above. They don't
make a lot of noise, or smoke, or smell like restaurants or bars," Toma explained. "And these basements are good for firms because the leases are
cheaper and they get to be located in a VIP neighborhood." When pushing
basement properties, Toma focused on the upscale niceties of historic Bucharest as much as the quality and design of a given unit's renovation. "Just
within the block of this [converted basement]," Toma boasted as we stood in
front of an interwar villa, "there are vegan restaurants and a café that roasts
its own beans. Other neighborhoods don't cater to this kind of lifestyle." Cast

in terms of authenticity, the basement level brought central Bucharest's history and contemporary culture of cool within reach of the creative classes.

Approximation

While historic, nineteenth-century buildings offered dramatic ruins ripe for retrofitting, developers of newer construction sought to commercialize storage basements by blurring the boundary between below and above, between basement and street level. Their strategy rested on the recognition that a certain segment of clients can only afford to buy underground units, even though people overwhelmingly prefer to live aboveground. New building developers attempted to entice certain buyers to move underground by designing units to look and feel aboveground. This makes sense given that what is called the ground level is often not a geological given but instead a production of modern engineering. City planners flatten hills, drain lakes, and fill swamp beds to create the effect of a smooth, continuous surface that is easier to govern as much as to navigate.[24] While other cities have shown the ground plane to be a naturally unstable and shifting ideal, developers in Bucharest have made it clear that whether a given unit is understood as being above- or belowground is increasingly a decision to be made by individual property developers.[25] The qualities of abovegroundness and belowgroundness have become design effects, and the former renders nominally subterranean spaces more commercially viable.

"In the zone where we are now," Dan, the owner of a large residential development firm that I will call Comfort LLC, explained to me in 2017 from his offices in central Bucharest, "you can only build to a maximum of seven stories."[26] Zoning laws meant to regulate building density restricted the square meterage of living space in new buildings, limiting the profits of construction projects in a moment of soaring land prices. However, as Dan and others were quick to note, basement levels were not factored into the calculation of living space. "Since I am only allowed to build so high," Dan continued, "I'm maximizing the profitability of my most expensive land, my more central locations, by digging down to make an extra floor of semi-recessed, basement units."[27] The strategy, Dan and other developers readily admitted, was tricky to pull off successfully. Basement-level apartments were more expensive to build because of the cost of excavation, and aesthetic and comfort-related concerns about natural lighting and humidity—as well as more

Figure 6. English garden concept, 2018. Photo by author.

practical fears about flooding—drove down the purchase price of basement units by an estimated 15 to 20 percent. "And because they are built in and around the city center, they are part of projects that target higher earners and also people of a higher culture," he explained, referring to professionals and the creative classes. Projects for the truly wealthy, Dan told me, reserve their basements for fitness and game rooms. Dan's project, by contrast, sought to appeal to people on a tighter budget but who nevertheless had, in his words, "greater expectations" than Bucharest's working classes.

To thread the needle of heightened production costs, lower returns, and discerning consumers with deep skepticism, Dan focused his attention upon the underground's aesthetics, manipulating the surface plane so that underground units looked and felt a part of the ground floor. "We designed an 'English garden' concept to address these buyers' concerns," Dan explained. "The mental resistance to the basement level is real, and so developers have to be willing to hire special designers to help the buyers see the beauty of these spaces, to appreciate the value of these units. It's about selecting the special flooring and finishes, staging the business or apartments, and landscaping the garden just right." Dan walked me through the basics of his "English garden" concept. While the configuration of the living space mirrors aboveground units, the concept utilizes excavation and retaining walls to produce small, private courtyards that are flush with the basement level. The gardens themselves receive direct sunlight and are landscaped with grass; ivy climbs up the retaining walls to provide a more natural look. The gardens are accessible via floor-to-ceiling glass doors that connect to a living or waiting room. As with aboveground units, additional windows appear throughout the unit, although they are positioned at sidewalk level, offering views of the shoes of pedestrians and tires of automobiles. "Even though the basement floor gets less natural light, even though they look out onto the sidewalk, you can focus buyers on the opportunity to step directly outdoors and into a garden. They'll find that it is quite okay—it will feel like they are on the ground floor."

Darius, who owned a firm with a portfolio smaller than Dan's but geared toward a higher price point, agreed. "We have pretentions about being innovators," Darius told me in 2016 as we sat in a conference room at his boutique firm's office. Like Comfort LLC, Darius's firm incorporated basement-level apartments into their projects to maximize their value. "We're working with particular construction materials designed for the basement level. Done the right way, there aren't drawbacks or disadvantages with basement apartments," Darius assured me. "We've developed basement units with larger windows that are beautifully done. I'd take a basement apartment—no problem."

In fashionable Dorobanți Square, the strategy for blurring below- and aboveground took different shape. Rather than excavating the ground level downward to create courtyards flush with the basement level, commercialized basements instead built upward to produce entryways that gave the impression that the business was at ground level. In 2019, an upscale brasserie, for example, operated out of a renovated basement in a historic villa.[28] To blur the boundary between below and above, the owners extended the villa's façade outward, enclosing the space between the original building and the sidewalk. Viewed from the street, the brasserie's entry door was framed by tinted bay windows that extended from foot to eye level, giving the impression that visitors will walk straight through, directly into the ground level. On the other side of the door, however, was a flight of steps leading visitors downward into the basement. At the bottom of the brasserie's stairs, visitors entered into an aesthetically repurposed underground. The basement cellar's brick walls were stylized with accent paint and white subway tiles, and the natural lighting from the entryway windows was enhanced with vintage

Figure 7. Basement brasserie conversion, 2017. (top) Exterior; (bottom) interior looking outward. Photos by author.

Edison light fixtures. Visitors settled into Scandinavian wood tables and chairs. For those caught up in conversation, the staged materiality generated the feeling of being at a fashionable café at street level. It was only really the solitary visitor looking to indulge in the pleasures of gazing out onto the world beyond, to take in the million little dramas of street life, who would be given pause.[29] Even the elevated perch of a bar stool left the café flaneur frustrated, staring out onto bicycle tires and calves encased in skinny jeans. Several cafés, restaurants, and salons along Dorobanți, and similarly upscale neighborhoods such as Amzei Square, adopted variations on this design strategy with commercial success.

Through an array of interventions, municipal and market actors aestheticized the Bucharest underground, materially restaging its Metro stations, basements, and cellars to generate new kinds of atmospheres underground, ones that would appeal to the dreams and aspirations of the middle classes. The diversification of aesthetic strategies speaks to the diversity of tastes and subjectivities within the middle classes, drawing the young creative on a night out and the entry-level data analyst at a corporate office party alike underground without requiring that the two intermingle. Those who developed and marketed these spaces put forward different appeals intended to overcome the hesitations, whether real or imagined, of those who have to go underground to live, work, or commute. As different segments are enticed beneath the city, taken with the convenience, coolness, or centrality of their new placement underground, the project of class separation across and within the city grows ever more refined. The underground's gentrification makes it possible to thin the presence of the middle classes on the streets of the city center without prohibiting access to these downtown neighborhoods. By extending downward, the underground provides an affordable foothold that facilitates dreams of working, living, and playing in the heart of Bucharest.

CHAPTER 2

Station Kiosks

Following the end of socialism in 1989 and through 2021, below the glossy storefronts of the multinational brands, boutique concept stores, and banks that had begun to take over Bucharest's city center, there ran a parallel track of commercial real estate. During that time, upward of 330 commercial kiosks had opened inside of the city's Metro stations.[1] Mostly simple, metal-framed stalls, the city center's second-tier shops facilitated cheap grab-and-go purchases for the Metro's 800,000 daily riders, at least until the ratcheting pressures of gentrification abruptly cleared the shops out entirely in 2021.[2] At the time of writing, most of these locations remained vacant as administrators considered what, after thirty years of operation, might take their place. "They're nothing special," Nandru admitted in 2015, when the fate of these kiosks was still open for debate. Nandru was a senior manager at Sindomet Comserv SRL, the company that managed the Metro's commercial properties.[3] "There are situations where six kiosks in a row all sell the same things: cigarettes and juice, cell phone credit, toiletries and underwear, pretzels," he rattled off. In sharp contrast to the storefronts atop of main squares, outside and local observers alike characterized these station kiosks as eyesores that evoked unfavorable comparisons with Istanbul rather than suggesting the desired aesthetic affinity with Paris or Brussels.[4] Local administrators hoped publicly that the wave of gentrification that had washed over the city center's main squares and boulevards would soon seep downward, expanding the presence of multinational brands such as McDonald's (see Chapter 1) to the underground. Nandru, however, saw things differently. From his perspective, logistical obstacles made the prospect of expanding underground a nonstarter for most multinationals. To address the aesthetic shortcomings of the Metro's commercial landscape, Nandru argued that

Figure 8. Kiosks, 2017. Photo by author.

the best option was to help the small businesses already there improve their storefronts rather than replace them with multinationals. "There's plenty of interest from multinationals in taking over these spaces," Nandru admitted. "Those kiosks sit on valuable property." Yet the turnover in Metro commerce had proven slow at best. "The problem is," Nandru continued matter-of-factly, "multinationals can't make it work underground. They just don't fit down there."[5]

The spatial mismatch that hampered the gentrification of the underground preserved (at least temporarily) a place for middle-class entrepreneurs with more modest needs and budgets to do business in a city center that was being taken over by global brands—albeit beneath them. In this way, the underground opened up a critical site in Bucharest for shaping the entrepreneurial spirit that Romania's transition toward a market-driven society required.[6] Take Ion and Boian as two distinct cases in point. Originally from the Republic of Moldova, Ion immigrated to Bucharest with plans of opening a string of shops selling candies produced by the Moldovan chocolatier Bucuria. Bucharest's unexpectedly high cost of storefronts, however, pressed Ion's entrepreneurial aspirations underground in search of smaller spaces with cheaper rents. He quickly established a handful of Bucuria kiosks that were brighter and more ambitious than the drab booths that administrators wished to remove (and eventually did), but did not come close to approximating the production values of a global brand. Boian, by contrast, turned toward the underground in search of new opportunity for his already lucrative Romanian café franchise 5 to Go. With deliberately styled cafés proliferating across Bucharest and beyond, 5 to Go conformed much more obviously than Bucuria to the aesthetics of the rising middle classes and to the aspirations of administrators, even though the brand lacked global recognition. 5 to Go's first sixty branches had all been located at aboveground storefronts; after beginning a foray underground in 2018 with four locations inside the Metro, Boian signaled his desire to further expand 5 to Go's footprint in the Metro to as many as twenty cafés.[7] Lucrative as his business plans may have been, Boian's ambitions were ultimately curbed by administrative discontent surrounding kiosks in general. At the time of writing, the so-called eyesores had been taken down, and the spaces they once occupied remained vacant.

With their low barrier to entry, station kiosks had enabled a wide proportion of the middle classes to establish their own business in the most exclusive squares of the capital. The opportunity for small-footprint retail in the city center's subterranean annexes shaped what it meant to be a middle-class entrepreneur in a post-socialist city being remade by foreign investment, both at the low end (Bucuria) and at the high end (5 to Go). This chapter toggles between the belowground efforts of Ion and Boian to pursue dreams of running a successful company in order to bring into focus the kind of entrepreneurial subject that the underground made possible, as well as the compromises and constraints to which these entrepreneurs had to adapt. While the intensifying forces of gentrification ultimately closed and then cleared out station kiosks en masse, these underground kiosks nevertheless spurred the development of entrepreneurial dispositions and sensibilities that would go on to find expression in the redevelopment of Bucharest's basements (Chapter 3) and cellars (Chapter 4).

Out of Reach

"No one factored in commercial projects into the original design of our Metro stations," Nandru explained to me back in his offices in 2015. "Of course, there were stores during communism. But they were on the surface—in blocks and in malls—but never in the Metro." When Sindomet opened stations for business shortly after the end of socialism, they carved out slivers of space for newly installed kiosks along the walls of already crowded Metro corridors and tucked them tightly into the corners of busy station galleries. The introduction of these improvised retail spaces opened up new opportunities for the city's small-time investors to become business owners. "When kiosks first opened, it was an untapped market. Everyone with some foreign money in their pockets thought about investing in one. The rate of return was huge," Nandru recalled.

The early excitement in investing underground is easy enough to appreciate. Besides being cheap to rent, the kiosks were also inexpensive to operate. The average-sized kiosk was small enough to be stocked and staffed by one or two employees. In the early days, there were few, if any, aesthetic expectations for staging a kiosk. With only minimal upfront costs, no-frills vendor stalls servicing a reliable flow of daily commuters promised big returns. Those who bought into that promise, however, soon realized the constraints of operating in the Metro. In practice, kiosk owners admitted time and again, the stalls proved to be as cumbersome as they were cheap to run. For example, kiosks were not connected to basic commercial infrastructure such as freight elevators to stock goods and remove trash, ATM machines to facilitate cash transactions, or restrooms for employees and customers. Sindomet also leased kiosks only on an annual cycle, discouraging businesses from investing in their stalls should their request for a renewal be denied. Finding success underground required a particular kind of entrepreneur, one who was agile, adaptable, content with a small footprint, and accepting of the precariousness that comes with short-term leases.

While the metal kiosks hawking convenience goods that took shape in the underground's particular business environment came to be regarded as eyesores, their ability to facilitate quick transactions that met immediate needs made them hugely popular with the Metro's ridership. As the Metro rushed passengers beneath overcrowded roadways, kiosks supplied the caffeinated beverages and sugary snacks that the city's new middle classes relied upon to keep up with the ever-accelerating demands of the globally competitive marketplace.[8] "The profile of the Metro is the middle class," Nandru insisted. In the absence of any collected information about the Metro's ridership, developers and businesses alike were left to feel out the underground's demographics for themselves. "They are people without a car, or if they have one, they don't want to burn the gas money sitting in traffic. And they find themselves caught up in this daily rat race." Nandru invited me to look out his window and onto the business district beyond as he spoke. "Office workers are riding to Pipera—every corporation seems to have an office there now—and they're all addicted to coffee and cigarettes. They also need to eat and drink. They're all in a rush. And they've come to rely on the Metro to keep them going." Nandru's assessment rang true of my own relationship to these kiosks. In addition to keeping me caffeinated, I had come to lean upon Metro kiosks for a remarkable range of needs in order to extend my days "in the field," from new batteries to keep my Dictaphone recording to toiletries for negotiating sharp transitions from observational work to office interviews.[9] "Metro riders don't have the largest salaries," Nandru admitted, "but there are hundreds of thousands of them rushing through the Metro every day, and they're collectively spending millions each year."

The tremendous flow of middle-class passengers with money to spend made the underground an intuitive site of investment for multinational brands, in many ways more so than the retail space along the streets above it. "The advantage of doing business in the Metro is the visibility," Nandru expanded. Analysts agreed. Even though Metro stations had only a marginal amount of commercial space to offer, they nevertheless hosted upward of 800,000 riders per day, a level of foot traffic that far outpaced Bucharest's most popular malls.[10] McDonald's had demonstrated how lucrative it could be to serve large volumes of customers, even if they only had small bills to spend (see Chapter 1). Metro administrators were eager to harness the Metro's commercial potential by having more established brands akin to McDonald's take over its commercial spaces. For example, the upmarket grocer Mega Image established a store in one of the system's most generous units. Administrators hoped that similar multinational companies would follow suit, and in the process improve the aging stations' overall quality and character, replacing metal stalls with aestheticized storefronts designed to resonate with an upwardly mobile middle class's wants and desires. But such a transformation, Nandru insisted, was easier said than done.

"I agree that modernizing the Metro is urgent," Nandru continued. Sindomet's contract with Metrorex (the company that administers the Metro system as a whole) to manage its commercial kiosks eventually lapsed because of its inability to improve the character of the system's commercial spaces. "Metrorex wants to throw existing shops out for multinationals like Carrefour and Auchan, but those kinds of stores require another kind of space entirely," Nandru explained. Lucrative as they may be, the available dimensions of the Metro's retail spaces, as well as the short terms of their

Figure 9. Grocery store, 2018. Photo by author.

leases did not fit the business model of multinational chains. "Ten square meters of floor space would be a big size for the Metro," Nandru continued. "But the business models of multinationals require a hundred or two hundred square meters of floor space." Achieving economies of scale and maintaining reliable inventory of a broad range of goods required more room, and generating a global brand's customer experience required longer-term leases to recoup the high upfront costs of staging a storefront than Metro kiosks could offer. Nandru rattled off a string of multinational brands that had expressed interest in operating inside Metro stations but were ultimately unwilling to operate within its constraints. "They come, they tour, we talk, and then I don't hear from them again," he admitted. "To bring in multinational brands to the Metro would require a fundamental rethinking of entire stations because multinational corporations want bigger spaces." The most daunting hitch, Nandru insisted, was not that stations were drab or that their finishes required updating. Those fixes would be simple enough. The problem was that the parameters of the spaces themselves were too small for the Metro's evolving ambitions and the spaces could not easily be expanded given the existing dimensions of stations. "A renovation that simply made commercial stations bigger wouldn't work because they would block evacuation routes and raise safety concerns," Nandru reasoned. Upgrading the Metro's commercial offerings, he speculated, would require the expansion of the stations themselves, which would be both difficult and costly because of their location underground. "And so, for now," he added, "locally owned kiosks are thriving." The needs of middle-class entrepreneurs, as represented by Ion's bootstrap candy shops and Boian's more polished café franchise detailed in what follows, proved to be very different from those of multinational retailers. The adaptability of these small businesses enabled them to squeeze within the constraints of the underground. Nandru pointed to Bucuria specifically as evidence that the Metro's existing small-scale entrepreneurs were absolutely capable of achieving the kind of aesthetic transformation of their kiosks that Metrorex demanded. "With a little more attention, we can change the face of these shops and improve the comfort of riders," Nandru continued optimistically. "The problem isn't the type of shop," he deflected. "The problem is that the stations themselves need to be entirely rethought."

The inability of global brands to fit within the Metro kept rents underground well below the city center's market average. Sindomet required kiosk owners to sign a confidentiality clause alongside their lease agreements, making it difficult to determine the exact gap between street level and underground rents.[11] Newspapers, however, placed the average monthly rent per square meter across the Metro system at between €8 and €10.[12] While declining to speak in specifics, Nandru admitted a wide variation in rental prices underground depending upon the centrality of the station, the kiosk's overall size, and its location within a station. At their highest, investigative reporting suggested, rents reached €51 per square meter in Union Square, €40 per square meter in Victoriei Square, and €39.60 per square meter in Aurel Vlaicu Station (which, like Pipera, serves the city's growing business district).[13] By contrast, street-level rents around University Square during the same time period averaged €65 per square meter.[14]

The specific space and temporality of the underground, ultimately, preserved an affordable foothold for middle-class entrepreneurs to open their own businesses in some of the wealthiest squares of the city. Though located underground, the Metro's commercial kiosks nevertheless created a narrow opportunity for local entrepreneurs to connect with the city center's relatively well-heeled population of office workers, government bureaucrats and administrators, and university students. Succeeding where multinationals failed to take hold, however, required an entrepreneurial subject attuned to the underground's constraints as much as its opportunities. "People generally start out with one kiosk," Nandru explained, offering his bird's eye view, "but I have clients now who have six, seven, even ten stores. They build out when the market is hot, and they know how to retrench in a crisis to avoid getting wiped out." As became clear over my conversations with Ion and Boian, the entrepreneurial subject found underground knows not only how to fit into the underground but also how to keep pace.[15]

Bucuria

"We didn't start in the Metro, but it's where our business is growing," Ion explained as we walked toward one of his candy kiosks inside Union Square Station. "We identified the first place, and then a second, third, fourth—now I have eleven stores," he proudly enumerated. Each of his kiosks sold Bucuria-branded chocolates and espresso drinks to go, beneath some of central Bucharest's busiest and most lucrative squares. Multinationals such as Starbucks, Paul's Bakery, and McDonald's occupied the critical real estate just above his kiosks. Up there, customers paid twice as much not only for their lattes and sweet treats, but also to enjoy the carefully staged atmospheres that these multinational corporations provided. Such firms vastly surpassed Ion's capacities in terms of the scale of his business and his ability to meet consumer

Figure 10. Bucuria kiosk, 2015. Photo by author.

expectations as to how space should be materially staged. "We first tried to get set up with street-level storefronts, but there was no way we could afford it—their rents were extraordinary. Even shopping centers like Cora and Carrefour were too much," he admitted. Not only was the cost per square meter higher aboveground, but the dimensions of the available properties were also significantly larger, driving the overall operating costs up even higher. "We would be paying more than we make to set up at street level," Ion said with a shake of his head. "I can only mark up the price of chocolates so much compared to what they are in Moldova," where the cost of living is considerably lower than in Bucharest. Rather than packing up and returning home, Ion instead pursued his entrepreneurial vision underground.

Compared to street-level storefronts, the Metro provided Bucuria with a lower bar of entry into the city center. The easier rents, smaller footprints, and minimal staffing of kiosks provided Ion with a space where he could turn a profit selling relatively inexpensive Moldovan goods in the heart of Bucharest. Ion quickly found, however, that the opportunity to establish not just one, but multiple, storefronts underground came with compromises. Having initially intended to operate at street level, Ion imagined his candy shops as being of a higher quality and character than a Metro kiosk. "Not [just] anyone can build a respectable business down here," he noted, drawing a comparison between Bucuria and other stalls nearby. "The aesthetics in the Metro are way off. Even if you succeed in making your own kiosk nice, you still have to deal with your neighbors." Ion wrinkled his nose at a kiosk catty-corner from his selling cheap lingerie dangling off of wire mesh racks; somewhat ironically, Ion's critical assessment of the Metro's general commercial offerings closely mirrored that of the administrators calling for the clearance of all kiosks to make room for multinational brands. Eager to set Bucuria apart from the rest, Ion gave some consideration to the design of his kiosks. At Union Square, for example, he framed his sales window with a candy-cane striped exterior capped by a white awning. Viewed head on, from the perspective of an ordering customer, the kiosk closely simulated a street-level storefront. Brightened by accent lighting, the awning displayed the brand name *Bucuria* in bubble letters alongside the year of the chocolatier's founding, 1946. The nostalgically old-fashioned design invoked the era in which Bucuria was established as well as, Ion hoped, affecting a sense of cheer (*bucuria*, after all, means "joy"). While unable to generate the kind of atmosphere that global brands achieve with their storefronts, Bucuria represented an acceptable upgrade from the simple metal cages that dominated the underground at that time. Ion insisted that, once set up, his kiosks were easy to run. "They're not complicated spaces to administer—it takes just a few people per store," he explained. When pressed about the particularities of operating underground, Ion pointed to difficulties in restocking. In the absence of service elevators, he and other vendors had to carry their goods down into the Metro by hand, only to carry their trash and packaging up and out in the middle of the night when the Metro was closed.

The simplicity of Metro kiosks, coupled with their low rents and short leases, afforded Ion an agility in expanding and quickly contracting his business that would not have been possible aboveground. Underground, Ion could afford to acquire, stage, and staff additional kiosks just as easily as he could quickly close established sites without sustaining much of a loss. "I have eleven stores, but things didn't go well at each of them. Only five are open now," Ion admitted. While he hoped to reopen a few of his closed stores, he planned to let the leases simply expire on others. "What's become clear is that every station has its specific character in its circulation," he explained to me. As Bucuria branched out to new stations, Ion refined his understanding of how his desired customers move through the city. "Our customer profile fits most clearly with riders headed north," toward the central business district, Ion explained. "Union Square, Victoriei Square—there are a lot of offices and government buildings there, and that's where we see a lot of our transactions. Our typical customer is an office worker in their thirties and is of middle to above average income." Particularly at rush hours, the Metro provided Ion's thriving stores with a steady stream of office workers looking for little indulgences to get through their days as well as last-minute gifts for their colleagues around the office. Attempts at tapping into stations busy with a different profile of rider, meanwhile, quickly shuttered.

For scrappy entrepreneurs like Ion, the dream of entrepreneurial success opened up by the underground outweighed its compromises. While he left Moldova with loftier ambitions of establishing a string of street-level candy shops in central Bucharest, Metro kiosks enabled Ion to approximate his vision. "By going underground, I'm still doing business in the best parts of town, in the busiest stations," he insisted toward the end of my tour. Rather than being priced out, Ion's flexibility and modest needs enabled him to squeeze into the underground's space and to leverage the temporality of its leasing structure. Ion's agile business model even allowed him to expand outward aggressively and to retrench quickly as needed. His ability to fit, if not thrive, where established brands could not served as a point of pride. "I hear of research saying that multinationals are willing to pay ten times the rent I do but that they can't make it work down here," he recounted with a grin. "Look what happened to Relay"—a French newsstand chain that operated mostly in airports and train stations. "They opened in the Metro and then left because they couldn't handle the setup." But, though proud of his ability to respond to the underground's restrictions, Ion nevertheless was aware of the clear ceiling to the kind of mobility that the underground made possible. While strategizing where in the Metro to try making his next move, Ion voiced no expectations of growing Bucuria upward in the future to occupy standalone storefronts in or along the main squares above his kiosks.

5 to Go

If market forces pressed Ion's entrepreneurial ambitions underground, Boian turned toward the underground only after having found broad success with his coffee franchise 5 to Go at street level. Launched in 2015, 5 to Go sold espresso-based drinks, served in Pop art-themed cups, out of Scandinavian-designed storefronts at a flat rate of five lei per item. The concept quickly developed a following among university students and newly minted office workers, allowing the franchise to expand within a few years to one hundred cafés in Bucharest and beyond.[16] "We're not yet as recognizable as Starbucks or Dunkin' Donuts, but our attention to design has helped us go viral," Boian, a founding partner of the franchise, explained to me in 2018 at its centrally located headquarters. In contrast to Bucuria's narrow margins and old-time, folksy appeal, 5 to Go trafficked handsomely in a sleek yet warm minimalism staged for social media. "My partner is an interior designer by training. We've been incredibly deliberate with our design," Boian continued. "We work with natural wood and metal to create an experience that looks beautiful anywhere, especially on Instagram."

At street level, 5 to Go franchises regularly took shape in small storefronts in order to focus on take-out customers. The Metro, with its small-scale commercial spaces and high foot traffic, emerged as a natural site for the company's ongoing expansion. As 5 to Go's presence in the Metro took root, however, Boian became increasingly aware of the added challenges and subtle indignities that came with tapping into the Metro's ridership. As the founder of an image-conscious brand, Boian was given pause by the Metro's

Figure 11. 5 to Go, 2018. Photo by author.

drab interior and the preponderance of image-oblivious kiosks immediately nearby. "It's a lot harder setting up down there—it takes a few weeks longer to get the design right in the Metro." While 5 to Go franchises fully renovated each of their storefronts, Boian noted that surface-level properties were typically already connected to utilities in ways that Metro kiosks were not. For his newly acquired kiosks, for example, Boian had to establish basic hookups to the water and sewer pipes running above him. Pumps to draw wastewater upward added to the upfront costs of getting established underground. While with extra time and effort Boian could enable a kiosk to make coffee like an aboveground storefront, he struggled to make his kiosks look and feel like one of his street-level cafés. Boian expressed anxiety about the quality of the kiosks' materials as well as the clunky joints and hinges that held them together. "We've tried to mask these kiosks as much as we can with our choice of materials so that they look more like those on the street," he lamented. "But it's just not right. And then the Metro has certain rules," Boian groaned. "We have to display certain paperwork, on bright yellow paper, prominently in our window. It's impossible to blend it in. Things like that are a little embarrassing for us given our design commitments."

In addition to certain aesthetic compromises from a brand that prides itself upon the meticulous staging of its customer experience, the move underground also required compromises to 5 to Go's established business model. "The franchise contract for 5 to Go is five years," Boian explained to me, "and we can easily find sites at street level that will offer us a five-year lease. But in the Metro, contracts are for only a year." While Bucuria's simpler kiosk setup turned the Metro's short-term lease structure into an asset, 5 to Go's higher upfront design costs made the same temporality a liability. In this sense, 5 to Go's needs more closely mirrored those of a multinational brand. "We're investing a couple thousand Euros into our kiosks, and then we have to hope that—God forbid—Metrorex doesn't change directors and decide to cancel our contract," Boian continued. Given the high rate of turnover among senior Metrorex officials, at times driven by accusations of corruption, Boian's concerns were not unfounded; and they ultimately proved astute.[17]

Beyond cost structure, 5 to Go had to adapt its customer experience to the rhythms of the Metro. Boian targeted his clientele by locating each storefront in milieus crowded with offices, universities, and housing blocks, but the 5 to Go model also sought to increase sales by cultivating personal relationships with customers. In this regard, 5 to Go set itself apart from larger operations like McDonald's that traffic in a high yield of anonymous transactions (see Chapter 1). Framing coffee as a routinized consumer good, Boian continued, "At most of our cafés, something like 80 percent of our customers are recurrent: they come every day, Monday through Friday, at the same time for their coffee break. We train baristas to know their customers' names and their orders when they walk in. It's about making friendly relationships." The temporality of foot traffic underground, however, lent itself to another kind of interaction entirely. The coming and going of trains released large waves of customers expecting quick turnarounds. "The Metro is very different," Boian continued, "because it's not about rush hour but 'rush moments': one train comes, fifty people walk by, there's a pause, and then another fifty come. The rush is constant and the faces are always different. There isn't the consistency." The steady rhythm of rush moments affected the pace of service as well as how baristas interfaced with their customers. Instead of attempting to establish rapport by memorizing names and orders, baristas underground focused on speed in an effort to turn unfamiliar customers around quickly before the next train arrived. Anonymity, rather than interpersonal connectivity, framed these transactions.

For Boian—albeit perhaps less so than for Ion—the underground's opportunities ultimately outweighed its many concessions. For one, the Metro's ridership closely aligned with 5 to Go's own customer profile. "They're eighteen to thirty-five," Boian laid out for me. "They're university students and they're faculty, corporate office workers and professionals. They're cool and creative rather than snobby [*oameni snobi*], and they like to escape from the office a couple of times per day to buy coffee from us." The Metro, Boian noted, regularly concentrated this segment of Bucharest's population in one place, connecting university students with their campuses in the center as well as young office workers with their offices in the North, every Monday through Friday. "At first we avoided the Metro because it was too busy—in the morning there can be four or five rows of people forming and it's annoying to push through to get your coffee," Boian observed. "So, I made a study of each entrance at University Station, tracking how many people come and go through each stairwell. We started with the station's quietest entryway, and the results have still been extraordinary." Much of the success, Boian remarked, had to do with the kind of purchasing trends that the underground's accelerated rhythm encouraged. "The Metro is so profitable for us," Boian continued, "because our sales there are overwhelmingly in coffee—our highest margin product. The way the Metro works positions us to sell predominantly high-margin coffee rather than low-margin pastries."

By extending business underground, Boian aimed to fuse the 5 to Go franchise with the Metro system as a way of becoming further integrated into the habits and routines of the city's growing middle classes. "We are interested in being at any Metro station where there are many faculties, office buildings, and residential developments: University, Pipera, Pacii Stations," Boian rattled off as examples of each. "We want to set up enough cafés within the Metro that customers buy 5 to Go when they enter the station, drink their coffee on the train, and then buy another cup on their way up and out to the office." At the time of our conversation, Boian's vision had begun to take shape. He had established a storefront next to the Popesti Leordeni Station, which served a large Comfort LLC development catering to entry-level employees (see Chapters 1 and 3), and Pipera, where many of them work. "We want to develop like this across the Metro system—we want a bit of every point in and around each station."

A few years later, with Metrorex's decision to evacuate the Metro's kiosks in 2021, Boian's particular hopes looked to have been dashed. Yet, in their thirty-year tenure spanning Romania's post-socialist era of liberalization, these kiosks played an important role in shaping middle-class entrepreneurship amid a city center that was fast being overtaken by global brands. Though it attracted a range of business people with varying levels of capital and success, the underground's distinctive temporality and spatiality required entrepreneurs to be agile, adaptable, content with a small footprint, and willing to operate with none of the security that comes with a long-term lease. For Ion, the opportunity afforded by the underground more than compensated for its limitations; Bucuria could not otherwise afford storefronts in the neighborhoods where its higher-earning customers lived and worked. For Boian, who had already established his business at street level, moving underground entailed a greater sense of compromise: having to accept shorter leases, diminished customer relationships, and the loss of total aesthetic control that he enjoyed at street level. At both the low (Bucuria) and high (5 to Go) ends, entrepreneurs had to adapt and change their visions to succeed in the face of the underground's demands. While they missed out on the flash and cachet that would come with operating along the high street, the ability to do business in the best and busiest parts of town—albeit beneath them— nevertheless sustained their aspirations. And as the chapters that follow detail, such an entrepreneurial sensibility would also shape the recycling of Bucharest's basements (Chapter 3) and cellars (Chapter 4).

CHAPTER 3

Basement Apartments

"We live in a really nice place," Liviu boasted in 2017 as he aerated his glass of wine. We sat at a bar down the road from his home. "I mean, compared with my friends and colleagues it's very nice." Rich in cultural capital, Liviu and his wife held advanced degrees from Ivy League universities in the United States. At the time, they worked for cultural institutions in the city that paid, by Romanian standards, solidly middle-income salaries. After years of saving, they bought the second floor of a villa in a desirable residential neighborhood of central Bucharest. This part of town was free of the highly functional and densely constructed socialist-era housing blocs that dominated much of Bucharest's streetscape. Instead, the historic neighborhoods near Aviatorilor, Dorobanți, and Victoriei Squares, for example, were made up of ornately designed two- and three-story houses with private gardens. Set behind wrought-iron gates and tree-lined sidewalks, these neighborhoods were where embassies, prominent families, and high-earning professionals chose to locate themselves. But these neighborhoods also quietly catered to an aspirational middle class. Behind the heavy-handed presence of boutique cafés and the conspicuous displays of luxury sedans, some residents in these neighborhoods had only a tenuous grip on their properties.

"But I have to admit," Liviu continued, with enthusiasm draining from his voice, "the costs of living here are killing us. By the time we add up all of our bills and expenses at the end of the month, our income just disappears." Rather than getting priced out of central Bucharest, and pressed out to a more affordable property in a less prestigious neighborhood, Liviu and his partner turned toward the underground to buttress their finances. Liviu followed a growing trend in central Bucharest of capitalizing upon their building's

Figure 12. Villa in Dorobanți, 2016. Photo by author.

generous storage cellar by converting it into a livable space for rent. Within a month of his posting an advertisement for their basement conversion, a young couple who were just striking out on their own happily agreed to rent it.

"Ceaușescu didn't make basement apartments," began Cătălin, a real estate agent working in central Bucharest. "Back then it was all storage boxes and utility pipes." Since 2000, Cătălin explained, two residential markets pitched at the middle classes have taken shape beneath the city: one is a rental market for DIY basement conversions in the city center's historic villas; the other is a market for purchasing basement-level apartments that have been designed from scratch as part of newly constructed residential blocs, typically along the urban periphery (see Chapter 1). The emergence of both markets mirrors similar trends in other cities where intensified development has driven middle-class and even affluent housing into the ground, such as in Washington, D.C., New York, Toronto, Vancouver, and London.[1] "There isn't a lot of availability in the existing residential market," observed Mitică, another real estate agent with an office in the city center. "And there isn't a lot more land for new buildings. What's aboveground has been fully exploited, and so renters and buyers are reimagining the basement level." Cătălin, Mitică, and other real estate agents insisted that there was no shortage of interest. "Pricing is a lot more attractive down there," continued

Cătălin. "Categorically, basements are cheaper than any other floor by about 30 percent."

This chapter explores how middle-class aspirations for a respectable life in an increasingly unaffordable city are pursued underground, in the city's underutilized basements and cellars. As it details, disparities in material conditions mean that different segments of the city's middle classes experience the turn to the underground in different ways: some have become landlords and others tenants; some are buyers and others renters. These differences manifest in a divergent set of attitudes and calculations. Some look underground in order to live in the city center rather than along its periphery; others to be in new construction rather than old construction; still others to live on their own rather than with roommates. Despite these differences, the initial optimism grounding each of these strategies—whether as landlords, renters, or buyers—eventually gave way to ambivalence and doubt as the underground's must and mildew inevitably seeped through the pleasant veneer of new renovations. In this respect, the gentrified underground takes on the affective tenor of what Lauren Berlant calls "cruel optimism."[2]

Flipping

The center-north of Bucharest contains some of the capital's most desirable neighborhoods. But, while these neighborhoods project status, their historic buildings absorb money. "As nice as it is to live here, it's still an old building," Liviu confided. He lived with his family on the second floor of a postwar building. "Nothing is energy efficient about the place, so the heating bills are awful." Old windows, high ceilings, and thin layers of insulation caused Liviu to hemorrhage money on utilities on top of the property's high regular maintenance demands. To be clear, Liviu's choice of housing was aspirational. He chose to live in the most respectable part of town, alongside high-earning professionals with whom he imagined having a "horizontal comradeship" owing to his elite education, despite the precarious budgeting this necessitated.[3] To tighten their tenuous hold on their housing (and on their sense of status), Liviu and his partner anchored their ambitions in the underground. "We decided to convert our cellar into an apartment to bring in some extra money each month," Liviu explained. Framing his strategy in vertical metaphors, he added, "It's what keeps our heads above water."

Figure 13. Converted cellar renovation, 2017. Photo by author.

The upkeep costs were all the more pronounced for the nineteenth-century villas that cluster near Dorobanți Street, a few blocks away from Liviu's home. There, Francise lived with her university-aged son in a stately two-story building that had remained in the family since its construction in 1895. "The property has seen a lot," Francise explained. "It was bombed in 1945, requisitioned by the Nazis and then by the Russians, and then nationalized by Ceaușescu. It also survived the earthquakes of the '40s and '70s." Francise counted out the historical traumas that the building had endured on her hand. In the Ceaușescu era, government officials divided the home into three units and assigned two additional families to move in. In the absence of ownership, each of the families had little incentive to reinvest in the building, particularly in the most economically austere moments of the 1980s. "It's an extraordinary structure to still be standing," Francise remarked with pride in both the building and her family's capacity to retain it over the years.[4] "But it's expensive to keep it that way. The house needs more than we have to give," she admitted. As housing-related bills outpaced Francise's pension, she worked with her son Toma, a university student studying economics, to convert their basement. "We noticed the trend with our neighbors was to go into real estate—to make a rentable basement that could be a source of revenue long term," Toma explained. "So, we did the same."

Both Liviu and Francise felt their family's claim to belonging in the city center slipping through their fingers. The growing impasse between the demands of older buildings and the rising cost of upkeep bore down upon them, threatening to press both families outward to a more peripheral neighborhood and down the social ladder. Both begrudgingly took on the role of landlord in order to firm up their grasps over their existing lives. Importantly, Liviu, Francise, and their neighbors stood in stark contrast to landlords of investment properties in major Western cities who take advantage of hyper-expensive real estate markets by rent-gouging younger professionals who lack the family wealth to buy their own place. With their basement conversions, both Liviu and Francise were attempting to preserve their present rather than ascend upward. The sort of tenants attracted to the kind of basement rentals offered by Liviu and Francise, meanwhile, were trying to stake their claim upon middle-class respectability. Rather than trying to hold on to their social position, the tenants understood themselves as finding their footing. As explained in what follows, they imagined a future with an upward trajectory—physically as much as socially.

"These converted basements are a new market, and they appeal only to a narrow range of young renters," Mitică, the real estate agent, explained as he toured me around a neighborhood with several listings. "Older Romanians would never live in a basement unless they were desperate. But younger Romanians who are just starting their careers and who don't yet have children are open to them." Conversations with a half dozen other real estate agents working in central Bucharest confirmed as much: the market for basement conversions attracted a particular profile. "It's the young ones that go for the basements," commented another agent, named Ion. "Their purchasing power hasn't caught up with their potential. They're open to renting a basement apartment because it's cheaper and it's central. The location makes it still fancy." Another real estate agent, named Elena, agreed. "With basements, the renters are—let's just go ahead and say 'hipsters:' they're young, creative, don't have a lot of money but still want to be in the ultra-center."

The renter profile influenced the kind of investment owners made in their basement conversions. "With basements, the margins are limited by the kinds of renter that they attract," explained Cătălin. "Owners generally avoid making big, exquisite apartments in their basements. The idea instead is to make simple apartments with affordable rents that will make some money from a cellar that otherwise generates no money." Aware from neighbors that the basement rental market skewed young, both Liviu and Francise and Toma organized their basement conversion strategies with college students in mind. "I figured students care most about being near campus and are more flexible about the rest," Toma reasoned. As a university student himself, Toma drew upon his own sensibilities as well as the insights of his friends. "Our home is in walking distance of four universities. We felt certain we could find renters regardless of how the basement turned out."

Both families sought to minimize their investment into their basements. "Our basement is about thirty-five square meters," Liviu continued, "and it was in rough shape: the stucco was falling down, it had dirt floors, and the air was damp." Liviu invested €5,500 of savings for materials and covered the rest of the conversion in sweat equity. He plastered and painted the walls and laid down an inexpensive floor himself. He only worked with a contractor to install the plumbing for the kitchen and the bathroom. While in a similar starting condition as Liviu's basement, Toma and Francise's cellar was considerably larger. They also lacked the skills to carry out the conversion on their own, which required a larger upfront investment. Francise estimated that they spent about €20,000 to transform their storage cellar into a three-bedroom apartment with its own kitchen and bath. In both instances, the cash-strapped landlords furnished their apartments with Ikea-knockoff furniture. While their basements held out the dream of living in the most desirable of locations, the hyper-functional, low-cost, and ultimately low-quality characters of these underground conversions were part of the compromise that their tenants would ultimately have to endure.

The high demand for housing in central Bucharest, real estate agents maintained, made the investment of a basement conversion relatively low risk. With near certainty, owners were able to find willing renters. The monthly revenue generated from a basement conversion, Liviu, Francise, and Toma explained, was transformative. At a moment when the average net income in Romania was €470, Francise and Toma collected €400 per month between their four tenants.[5] The cash influx enabled Francise to keep up with her home. "The monthly rent goes mostly toward our basic bills, like heating and electricity," Toma explained. "About 20 percent is left over each month, and that is saved to cover home maintenance projects during the summer." Liviu's converted basement performed similarly. "It was always intended to cover our basic expenses," he explained. "To determine the rent, I added up all of our monthly utility bills and divided the number by twelve." When fully rented, converted basements generate just enough added income to preserve historic buildings with architectural character as well as to maintain a social position in the right part of town.

Although such rentals were increasingly common, the size of Bucharest's basement conversion market was difficult to estimate. Basement apartments, officials and owners alike agreed, fell largely beneath legal regulation.[6] The municipal government did not have a required permitting process to convert a storage basement into an apartment. To be sure, the state closely registered the extent, value, and ownership of a building, via the Cadaster (*Cartea funciară* or *Cadastru*). Assessing a property's value required differentiating between livable, technical, and storage spaces. The Cadaster, however, did not police how a building's space was actually used. "If you want to change a storage basement into an apartment," explained Vlad, a senior official for the National Agency for Cadaster and Real Estate Advertising, "the regulations are not very clear. If you decide you want to live in your basement, that's your choice. It's not up to us to check up on you and say, 'why are you living down there—this is supposed to be storage.'" The incentive to legally document a basement conversion only arose when an owner was ready to sell their property. Only at this point did it become beneficial to increase the official size of

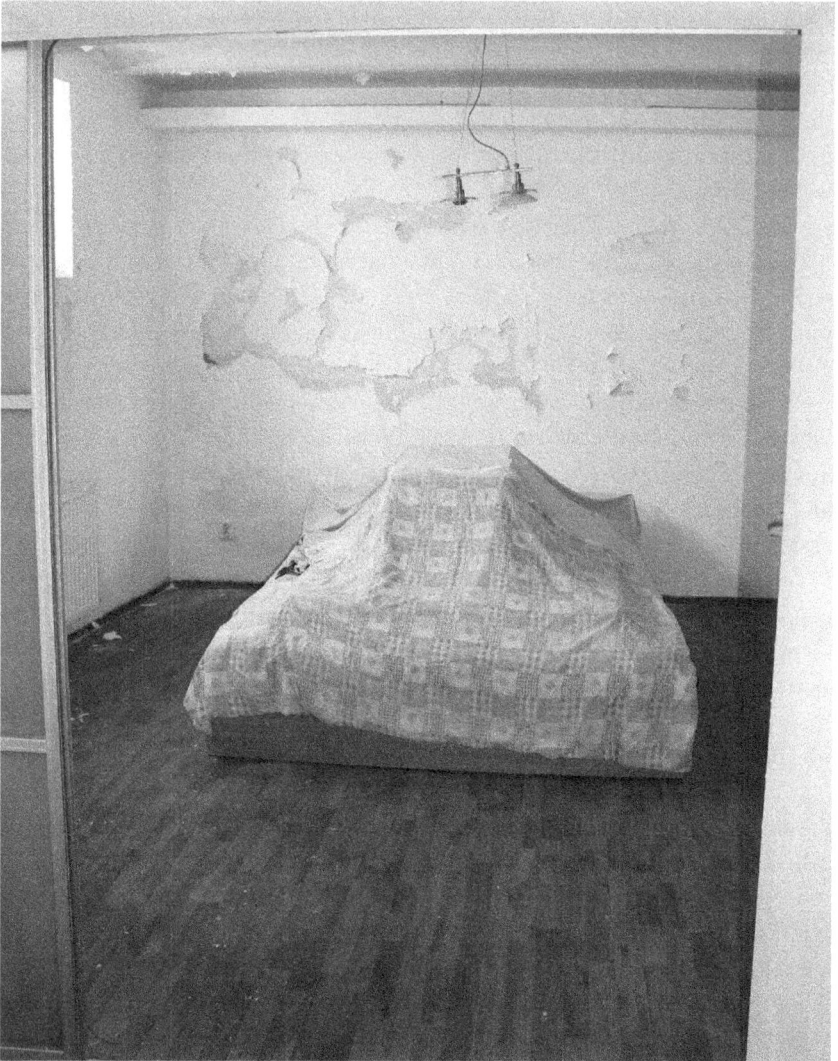

Figure 14. Water damage, 2015. Photo by author.

a property's livable space. "Otherwise," Vlad explained, "when you sell the property, you can only sell as it is written in our papers." Owners, eager to extract the maximum value out of their investments, tended not to register their rental agreements with the state to avoid paying taxes. The unregistered and untaxed nature of basement conversions contributed to their being priced at below-market rents. "Walk around the neighborhood," Francise suggested. "You can see apartments and even stores working out of the basement level. It's happening everywhere but no one's talking about it." With a chuckle, she added, "It's all under the table."

Although basement conversions helped to keep households afloat, the money was not easy. Once pressured into the role of landlord, overseeing and maintaining the basement rental required an unexpected amount of labor and expenses. Serving as landlord became a source of its own anxieties. Francise and Toma, whose basement conversion was older than Liviu's, were clear on this point. "Living in the basement is not like the other floors," Toma insisted. "There are some tricks to it." Although fully renovated, the air in Francise and Toma's cellar remained damp, and the humidity level only worsened with showering and cooking. Unless actively managed on a daily basis, condensation collected on the walls, causing the paint and masonry to flake off onto the floor. "You have to train each tenant on how to take care of the place, on how to live there," Francise explained. To keep the air dry, Francise asked her tenants to empty a humidifier twice a day and to open the cellar's three-inch windows for an hour in the morning and in the evening. The windows were located along the cellar's ceiling, leaving tenants without much of a view, and they were at foot-level with the outside patio. When left open, dirt, leaves, and twigs regularly blew down into the apartment. Unsurprisingly, tenants did not stick to the routine. Especially during Bucharest's frigid winters, tenants opted to deal with the mustiness of the cellar rather than the combination of blustery winds and blown-in detritus. Without fail, the walls were water damaged by the spring, the paint and plaster peeling off and falling to the floor. "Over and over again we end up having to redo the walls— the masonry and the painting," Toma lamented. "The basement has a lot of its own expenses. Taking care of it can be its own nightmare. But we're still coming out ahead with it." Beyond the constant cost and effort of keeping the basement conversion rentable, Francise also faced the difficulty of living with, and relying upon, university students within her property. "Last year I had four students living in the apartment, and every month they had a problem with rent: one could pay at the start of the month, another in the middle, and another toward the end. But the bills I pay with their rent money come all at once, not scattered across the month." In addition to tensions around cashflow, Francise noted that, although tenants live in the basement, their presence reverberated throughout the house. Their guests walked through the property's central patio, their noise carried up from the basement into the living room, and they passed through the kitchen to get to the laundry machine. Initially imagining the basement as a separate and distinct apartment, Francise had come to understand that, in practice, her underground tenants were now a constant presence in her household.

Although riddled with tiny indignities, the turn to the underground succeeded in anchoring landlords and tenants alike with the appearance of respectability. For both Liviu and Francise, their basement apartments facilitated a life in the city center's most exclusive neighborhoods, helping them to avoid being cast outward (toward more affordable neighborhoods) as well as downward (to the lower rungs of the social ladder) by economic pressure.[7] "The arrangement was good for me, and it's good for the renters," Liviu insisted. "We all get to live in what is the finest area of Bucharest! The Metro is close by, and one can easily walk to the universities and to the bars in Old Town at night." Pitched as a selling point for his renters, the close proximity to the city center's amenities was enjoyed equally by Liviu. Francise, by contrast, emphasized the improved quality of life afforded to her tenants by her villa as opposed to an apartment in a residential bloc. Her basement, she insisted, offered more space, fewer neighbors, and a private garden where they could entertain friends. "Life here is just better," Francise assured me.

Strategic Claims

Those who already had housing turned to the underground as landlords in an effort to preserve their social positioning into the present; the rest of this chapter, though, turns its attention to how those in need of housing moved underground. Some went underground as renters of the kinds of basement conversions detailed previously, while others became buyers of basement apartments that have been formally designed into newly built residential blocs along the periphery (see Chapter 1).[8] "In my fifteen years of experience in real estate, I can't recall a single buyer that didn't spend their entire budget," Cătălin summarized. "The mentality here is not to hold back money in order to refurbish. It's to get the best possible place they can afford in that moment."

The proliferation of basement apartments, Cătălin explained, has added new depth to the calculations of both buyers and renters.[9] Adding a vertical dimension to the horizontal considerations of cost and neighborhood that have long dominated the market, basement apartments opened up to the middle classes new room for maneuverability in the city, socially as much as physically. Basement apartments not only started with lower asking prices, Cătălin and Mitică explained to me, but also exhibited greater fluctuation during the negotiation stage. "Certain buyers are now asking themselves, is it better to live in Dorobanți [the chic neighborhood in the city center discussed previously] underground or in Obor [a traditionally working-class neighborhood] aboveground," noted Mitică. "Buying into Old Town in any way, at any level, allows you to interact with people of a certain quality," he continued. "You cannot have very high expectations for your neighbors in Obor." Throughout our conversation, Mitică pushed the kind of respectability that can be found underground, beneath the city center, while foregrounding the compromises that came with living in traditionally working-class neighborhoods. Similar to how the verticality of penthouses factors into the considerations of the truly elite, the subterranean extension of buildings has

Figure 15. Basement apartment unit, 2018. (top) View from outside. (next page) View from inside. Photos by author.

Figure 15 (*continued*)

entered into the calculations of middle-class buyers. "It's facilitating a certain lifestyle," Cătălin insisted. Conversations with those living in basement apartments made it clear that the decision to move physically underground was a part of a larger project geared toward assembling an upwardly mobile sense of self, not only in the present but also into the future.

Having One's Own Place

Speaking with me at a café in Old Town in 2017, Mădălina was in her early twenties and worked in an office in Bucharest's center-north. A few years prior, while a university student, Mădălina had opted to rent a basement apartment rather than live for free in a university dormitory. "The government offers free housing in the dorms for students like me," Mădălina explained, referring with modesty to a scholarship that she had obtained. "But the dorms would have been really hard for me—I need my own space. I don't like to share my things," she admitted with a self-deprecating smile. Instead of being subjected to communal living in the dorms, Mădălina turned to the underground as an inexpensive site where she could rent her own place. After a few weeks of searching, she found an inexpensive basement apartment rental in a newly built residential block next to the Titan Metro stop. "When you're a student you don't really have too much money," Mădălina conceded, "so it was difficult to turn down a free dorm room. But I need my own space, my privacy," she insisted. "That apartment was good for me because it gave me control—complete control—over my own space. I got to decorate it," Mădălina continued, "and I had my very own bathroom."

The inexpensive basement rental afforded Mădălina both freedom from the communal experience of student living as well as a space to experiment styling her adult self, one marked by the pleasures of arranging private property.[10] "I was the first occupant," Mădălina went on. "It was a two-room apartment too, not a studio." The sense of adulthood that came with having her own apartment resonated amongst her friends, Mădălina reported, many of whom lived either in the dorms or with their parents. "I was twenty years old. I was spending ten to twelve hours a day at the university anyway, so I was really just sleeping there. And instead of having to go back to a dorm, I got to go back to my own place," she explained. Waxing poetically, she continued, "At first it was like I had transposed myself into it. I made up that space so that it represented my personality," she beamed. Although it was an inexpensive basement apartment with equally cheap furniture, the newness of it all echoed Mădălina's own narrative about starting adulthood and building a career from scratch. The underground presented itself to Mădălina as an aesthetic manifestation of her arrival into adulthood—at least at first.

"Eventually, though, I found out about the disadvantages," Mădălina gave a dramatic eye roll and laughed to herself before rattling off the ignominies that came with establishing oneself underground. "The place was always messy," Mădălina recalled, "because if you left your window open a lot of dirt and dust would blow in. It was also dark all the time. There was a parking lot in front of my window, and there would always be a car blocking my light. When it was about to rain," Mădălina continued, "the place would get completely dark, like it was night inside. . . . It was cool telling people that I had my own place, but the truth is that I was ashamed to invite people over. I didn't want people to see how dark and dirty it was." The material conditions of the apartment gave Mădălina twinges of anxiety. "Sometimes I felt like I was in prison, because it's in a basement. I started eating my meals outside. I couldn't live that way anymore."

After graduating with her degree and accepting her first job offer, Mădălina immediately moved out of her basement and into an apartment that was both aboveground and more centrally located. "That first apartment was an important step for me," she reflected. "Back then, my family supported me financially, and so we agreed that a cheap place was a good place," she explained. "But now I'm grown up, and work is going well. My needs are nicer and my finances are greater," she smiled. In an unexpected demonstration to me of her rising capacity, Mădălina upended the conventions that come with having weathered the anthropologist's wearisome questions by picking up the check after our conversation came to a close.[11] Mădălina's basement apartment, in the end, provided that initial foothold to support her sense of ascendency.

Owning, Not Renting

For Ştefan and Viva, the basement level brought the dream of homeownership within reach, facilitating their own project of upward mobility. The couple purchased their basement apartment in a Comfort LLC building on the southern edge of the capital, a few minutes' walk from the Dimitrie Leonida Metro stop (see Chapter 1). When we spoke in 2018, Viva was staying at home to care for their two elementary-school-aged children while Ștefan pursued a career in the military. The move to Bucharest's southern edge brought Ștefan within driving distance of his job while still keeping the city's amenities within reach for their children. Both Ștefan and Viva placed a premium upon having been born and raised in central Bucharest. "We're second generation," Viva explained. "Our parents came in the '60s, during our Communism's industrial revolution." She referred to socialist-era policies of forced urbanization that had helped to transform Bucharest into a center of socialist industry (see Introduction). With an air of distinction, Viva added, "We both grew up in the city, close to the center." High rates of home ownership following the end of socialism in Romania left Ștefan and Viva with the expectation of owning their own apartment as well. The high cost of housing in Bucharest relative to Ștefan's military wages, however, made renting (much less owning) a difficult goal to attain. "I don't see rents lower than €300," Ștefan explained to me, "and that's really the most we could afford." Having grown up near the city center, both Viva and Ștefan placed a premium on remaining proximate to it. However, they could not afford to rent enough space for a family of four. "We saw about five or six apartments—we could have rented a smaller apartment closer to the center," Viva emphasized. "But when you don't have enough room, you can be living in a luxury hotel and still feel miserable," she assessed. Ultimately, they found their basement apartment on the southern edge of the city. "It was the biggest place we saw, and we could actually afford to buy it," Ștefan added.

As with Mădălina, Ștefan and Viva described their turn toward the underground as a kind of calculation, one that balanced the respective benefits of home ownership and renting, apartment size and proximity to the city center, residing above- and belowground. "We took everything into account before buying," Viva assured me: "the humidity, the lack of light, the smell." She pointed to her kitchen window, which looked onto a parked car's tire. When left open, the kitchen quickly took on the scent of car exhaust. "But we also looked at the good side. The apartment is well made. We don't have moisture problems, and there are plenty of parks nearby for the kids." Their mortgage proved, ultimately, to be cheaper than renting closer to the city center. "We save about €100 a month by buying here. And we can use that money to make life better in other ways," Ștefan explained. While Viva spoke of small treats added to the grocery cart and the occasional weekend trip to the seaside, Ștefan emphasized their desire to pay off their mortgage early. "Our goal is to constantly cut our expenses," he insisted. "We want to pay

our mortgage off in half the time. We see this place as a compromise situation. We're compromising by living here now so that we can move somewhere
better in five years."

Both Ştefan and Viva imagined their basement apartment as a vehicle for
their future upward trajectory in two senses: to move physically upward, into
an apartment fully aboveground, but also up the social ladder, into a more
respectable neighborhood closer to the city center. "I don't mind owning a
basement apartment that much," Viva opened up to me, "but I do mind living out here, on the periphery." Both Viva and Ştefan grounded their sense
of respectability in having been born and raised in the capital, in relative
proximity to the city center. Their present neighborhood, although within
the city limits of Bucharest, attracted migrants newly relocated to the capital
from the provinces. Trafficking in the language of incivility, Ştefan continued, "go check out the license plates of our neighbors." He pointed to the car
tire outside his window. "They're from Suceava, Argeş, Buzău," he continued,
rattling off a string of smaller, provincial regions. "They're uncivilized. They're
not properly educated—they don't yet understand how to behave in society,"
Ştefan said dismissively. To substantiate his assessment, he pointed to regularly observed behaviors that are commonly associated with the provinces:
spitting sunflower seeds, late-night drinking sessions outdoors, and smok-
ing in stairwells.[12] In sharp contrast with Mădălina, the major anxiety of
going underground for Ştefan and Viva was not the material conditions of
their apartment. Instead, it was a negative relationality with neighbors and a
perceived lack of familial autonomy. "But as I said—this is all temporary,"
Ştefan insisted, echoing Mădălina's orientation toward the future. "In five
years, we'll be able to buy into an old block two or three stations closer to the
center." Ştefan and Viva imagined their basement apartment as a temporary
compromise that would allow them to push upward into an aboveground
apartment, and inward closer to the city center.

New Instead of Old Construction

For Iulia and Geo, the basement level opened up the opportunity to live in
new rather than old construction. Iulia and Geo met while at university, and
they were now professionals in their mid-twenties. Iulia recently had bought
a two-room "English court" (see Chapter 1) unit in a new building located
next to the Păcii Metro stop along Bucharest's western edge, while Geo was

renting a similarly sized English court apartment in a nearby complex. Both women used the Metro to commute to their office jobs in Bucharest's center-north. Neither Iulia nor Geo initially intended to live in basement apartments in Militari; they had focused their early searches in centrally located neighborhoods around where Liviu and Francise lived and rented out space. However, Iulia and Geo were quickly turned off by the high premium expected for aging buildings in desperate need of renovation. "In Bucharest, a lot of owners in the center expect to get a lot of rent just because their place is in the center, and so they never invest in the place," Geo insisted when we spoke in 2016. With a wrinkled brow and scrunched nose, she described unrenovated studios in aged buildings. "I mean it was a tiny room with a dirty floor, and it had the shower set up above the toilet. It felt like rats lived there."[13] Despite the dirty and cramped conditions, the listed rent exceeded her maximum budget of €350 per month. Iulia nodded knowingly before speaking of her own disappointments with the high cost of old construction near the center. Real estate agents indicated that Geo and Iulia's calculations were common and indeed were starting to shape the residential market. "The unwillingness to invest in the repair of older buildings," Mitică affirmed, "has brought about a preference for newer ones." His colleague Marius agreed. "Basement apartments become particularly interesting investments because they provide a cheap way into a new building, often with more space and for less money than a studio in an older block or villa."

In their separate searches, Iulia and Geo each turned (albeit reluctantly) toward basement apartments in order to bring new construction with sleek modern finishes within reach. Although hesitant, Iulia was ultimately won over by her building's use of enlarged windows and a recessed garden to effectively blur the distinction between above- and belowground (see Chapter 1). "When I saw the ad, I didn't even realize it was in the basement because of the garden and because of the ceiling-to-floor windows in the living room," Iulia explained. The long and narrow skylight windows that line Bucharest's basement apartments appeared on her unit's other walls. "I was skeptical about living in the basement level," she admitted, "but the apartment was €10,000 cheaper than anything else I toured. The money I saved allowed me to buy a car and to hire an interior designer. And now I love the place." The underground extended to Iulia not just the dream of home and car ownership, further reinforcing her sense of having established herself as a young professional, but also the pleasures of meticulously personalizing her private living space.

Geo, whose English court rental was located next to Iulia's building, described a similar excitement upon moving in. "I didn't search for basement apartments at first because I had the usual prejudices: that it would be dark and damp and dingy," she began, "but it was the nicest place that I could afford to rent." Like Iulia, Geo noted feeling at home in her apartment. "It's comfortable—I don't feel underground like damp or cold," she insisted. Over time, however, the practical realties of basement living took a toll on Geo's enthusiasm. "I totally see, though, that I'm underground. Outside my windows are car bumpers and sneakers [Adidas]. I started keeping my blinds closed because I know people passing can actually see into my apartment, like what I'm cooking or where I'm sleeping."

For both Iulia and Geo, the main anxiety associated with living underground has been the difficulty they have experienced convincing their friends and colleagues of its benefits. "Whenever I tell people that I'm living in a basement unit, they start looking at me funny, like I'm poor," Iulia recounted. Geo nodded empathetically. "But I'm not actually that poor, I just like the apartment: it's new and clean. I'm the first person to live there. I know it sounds crazy, but I felt right away that I could make the place my own—that I could imprint my vibe on the walls," Iulia explained, echoing Mădălina's sense that the newness of her basement facilitated an aesthetic articulation of her new life stage and status as a bourgeoning professional, even if she could not convince her friends and colleagues to see the excitement. With newness also came a sense of certainty in the construction of the building itself—that the outdoor elements would not seep inside, that pipes would hold rather than burst, and that the building would stand through an earthquake. "I've been living there a year," Geo added breezily, "and I don't have to worry about a thing because the building's new. I wouldn't feel the same in an older building." Instead of fretting about mold and dust, both Iulia and Geo took comfort in the pleasures of their gardens, gently bragging about the luxuries of sipping their morning coffee while bathing in the morning light and lounging outdoors with their friends in the evening, drinking wine.

Although their complexes are located on the edge of the city, neither Iulia nor Geo expressed status anxiety about having a peripheral address as Ştefan and Viva did. Whereas Ştefan and Viva articulated concerns that the capital's periphery was bleeding too closely into the provinces, Geo and Iulia understood themselves as still connected to the city center by way of the Metro. They also imagined themselves as ascending with the neighborhood. "The fact is that a lot of investors are building things here [in Militari]. In the next five years I think it's going to be a really modern neighborhood. Out here is where the new and exciting stuff will be happening," Iulia speculated optimistically. In the meanwhile, both Iulia and Geo distracted themselves with Old Town's already developed leisure economy. "In one sense I've never lived further from Old Town than right now," Iulia noted. "That's one of the reasons why I was hoping to find a place in the center. But," she shrugged, "when I get bored I can still get there in under fifteen minutes. Metro access is everything. . . . Unless of course it's the end of the night," she joked, glibly referring to the fact that the Metro closes before bars and clubs do (see Chapter 4).

*   *   *

This chapter explored a myriad of ways different segments of the middle classes turned to the underground to make a claim to middle-class respectability in a moment of mounting economic pressure. Important material differences shaped how the middle classes utilized the underground, whether as a landlord or a tenant, a buyer or a renter. These differences also shaped the temporal registers in which they operated: in some instances, to preserve a past status into the present; in others, as a temporary inconvenience to bring a desired future within reach. In all these cases, though, the underground played a key part of aspirations to live well in an increasingly unaffordable city. The affective arcs these different middle-class residents experience are strikingly similar too: the initial optimism each found in being in the city degraded toward ambivalence as the constraints of living underground became apparent.

CHAPTER 4

Night Clubs

"The scene here is nuts," John, an actor from Los Angeles, California, marveled. John had come to Eastern Europe for the summer of 2018 to let loose after his character was killed off on a popular HBO series. He had heard about Bucharest's nightlife district, Old Town, while partying his way through Budapest and Berlin and had reworked his itinerary to experience firsthand what all the fuss was about. John offered me his unsolicited five-star review after I obliged his blind request for help ordering from a menu at a restaurant outside of the Old Town quarter. "Around 4 a.m. I made my way into an underground club, and there was no sign that the party was letting up," his recap of the night before continued. As the capital's only pedestrian zone, Old Town attracted more daily visitors to its three hundred bars, clubs, and restaurants than any other entertainment district in Bucharest, generating tens of millions of euros annually in revenue.[1] The neighborhood's ability to attract foreign tourists like John evidenced the far-reaching success of its revitalization. The Old Town development had helped Bucharest shed its post-socialist reputation for poverty and dilapidation and rebrand itself as Europe's newest party destination.[2] Old Town's ability to attract an international crowd night after night lent the neighborhood a cosmopolitan atmosphere that only heightened its appeal among local young professionals and college students. "The underground club was packed. When I came up, the streets were still packed. I've never seen anything like it," John gushed from behind his aviator glasses.

Mihai, the manager of Old Town's most storied nightclub, Club A, testified to the neighborhood's sudden transformation into a successful nightlife district.[3] "When I started working here in 2006, the neighborhood was something else," Mihai recalled when we spoke in the summer of 2015. We

Figure 16. Club A, 2016. Photo by author.

were standing atop Club A's entry stairwell watching a constant stream of revelers spilling past while we chatted about the changing character of the neighborhood. Founded as a nightclub by and for architecture students at the nearby university, Club A opened in 1969 in an abandoned cellar. By 2015, it served as a cultural point of reference, if not an outright model, for the dozens of other subterranean bars, clubs, and cafés located in basements and cellars that animated the district. "I mean, ten years ago there was no historic center to speak of. This area was completely abandoned. There were some 'gypsy' (*țigani*) families squatting around here but otherwise . . ." Mihai gave an introspective chortle before sucking down the rest of his mojito through a thick plastic straw. The muffled sounds of revelry could be made out coming from below, and similar sounds poured out of the other street-level bars and clubs located around us. "I mean, this place being called Old Town is only about five years old."

The Old Town development is what remained of the historic center following the efforts of the former dictator Nicolae Ceaușescu to remake the capital into a monument to Romanian socialism.[4] Once the home of artisans and craftsmen, back when Bucharest enjoyed the reputation across Europe as the "Paris of the Balkans," Old Town had fallen into disrepair after the 1977 earthquake. Following the end of socialism, and in anticipation of EU accession, the largely pre-twentieth-century building stock caught the attention of investors interested in developing a nightlife district in Bucharest.[5] The municipality supported this effort at neighborhood revitalization as the renewed area around Strada Lipscani (Old Town's central artery) would serve as material proof of Bucharest's European history.[6] With the backing of the European Bank for Reconstruction and Development, as well as local investors, Old Town's main corridor began gentrification in 2005.[7] As part of the process, Roma families were brutally evicted, clearing the way for what others have described as a "commercialization tsunami."[8] Within a few years, Old Town quickly transformed from derelict to chic, coming to play a central role in the nights and weekends of Bucharest's young middle classes while also attracting thousands of foreign tourists.[9]

This chapter explores Old Town's dramatic underground-up transformation. It examines the vertical development of the neighborhood's nightlife district, which began inside its historic basements and cellars and then expanded upward. These underground spaces facilitated novel practices of vertical segmentation that opened up new temporalities of investment, consumerism, and revelry, which incorporated Bucharest's growing middle classes into the very heart of the city center.[10] Yet as the neighborhood matured, the risks of incorporation by way of the underground became apparent. Underground spaces proved distinctly susceptible to the bureaucratic corruption and pursuit of quick profits that framed business generally in Romania at that time, ultimately compromising the safety of Old Town's middle-class clientele.

Sub- and Superstructures

At the turn of the millennium, Bucharest's historic center was "completely destroyed," "abandoned," and "falling apart," according to the initial wave of club owners when we spoke in the summer of 2015. The facades of buildings were cracked and crumbling, their windows lacked panes, and their roofs threatened to cave in. "When we got started," recalled Corneliu, who in 1999 opened one of the district's first clubs in the cellar of a dilapidated villa, "there were some shops in Old Town selling odd things, like shoes and cheap wedding dresses." Nelu, who opened his basement club in 2007, provided a simi-

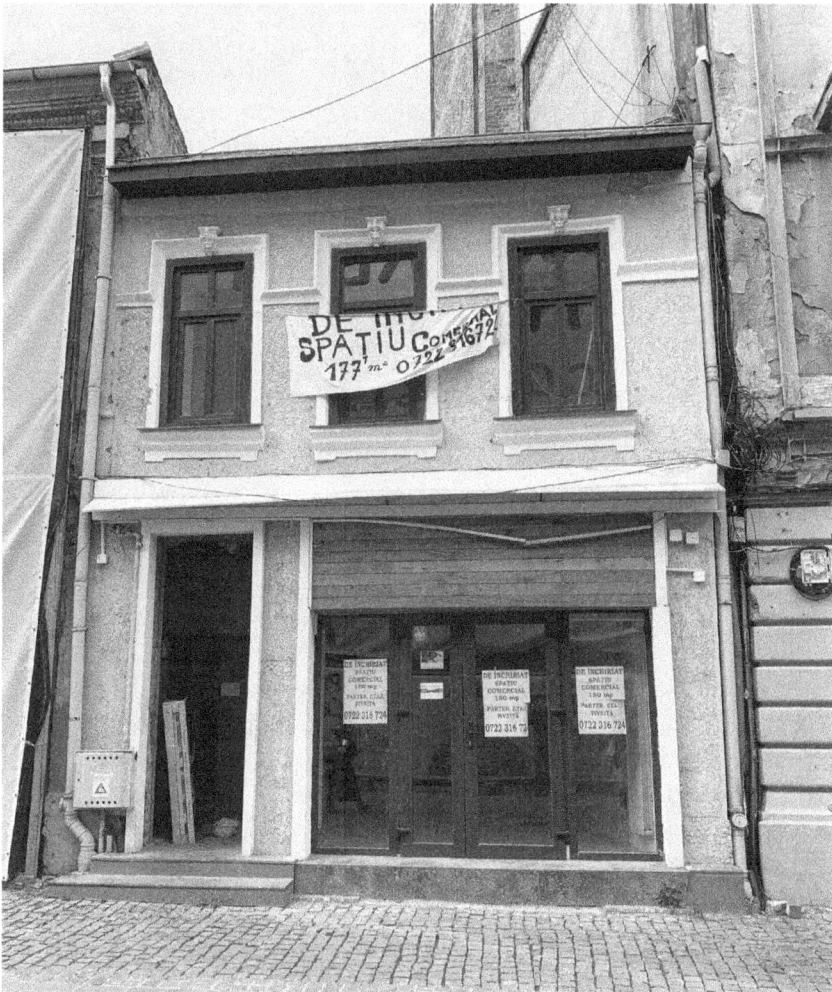

Figure 17. Newly renovated building with cellar for sale in Old Town, 2019. Photo by author.

lar assessment. "It was nothing like today," he explained. "We took over a falling-apart building that had been used as a general repair shop fixing sewing machines and vacuum cleaners." Mihai, Nelu, and Corneliu all took part in a late-night lifestyle, but behind the tousled hair and graphic t-shirts, each of these owners understood themselves to be business professionals, and they operated as such. They discussed Old Town's nighttime economy in terms of

Figure 18. Converted brick cellar, 2015. Photo by author.

"entertainment products," market niches, and target audiences. They also regularly emphasized the need to move enough drinks each weekend to make
payroll and to clear their bank loans at the end of the month. While a multi-
tude of Romanian students, professionals, and foreign tourists headed to Old
Town looking to let loose, the investors who owned the venues approached
things more strategically. Affecting a sense of market command, both Nelu
and Corneliu were clear that there were many other available properties across
Bucharest, most in better condition, that could have accommodated their
nightclubs. Corneliu, Nelu, and others, however, speculated that Old Town's
historic structures conveyed a future potential that made the neighborhood a
more intuitive place to invest compared with other zones in the city. "In every
capital in Europe, the historic center is a place to go," continued Nelu. "You'll
find bars and clubs, cafés, restaurants there. We knew one day this area would
take off in the same way. So, when we opened our club, we were investing in
the neighborhood's future potential." This makes sense given that at the time
it was widely perceived as being in ruin. "And of course," Nelu continued as
though it was self-evident, "we started underground."

 If the neighborhood's historic superstructures conveyed its future potential as a nighttime destination, Old Town's initial bars and clubs took shape

underground, inside the substructures of its basements and cellars. The dilapidated superstructures, charming as their future potential may have been, were too costly to refurbish. "Renovating a historic building is a very delicate thing," Sorin, a Romanian developer (see Chapter 1), explained from his offices in central Bucharest. He offered a comprehensive analysis of the Old Town development. "Some of these buildings are monuments, and most all of them are at risk of collapse. From the perspective of costs and permits," Sorin continued, "it makes no sense for these small businesses to take on the risk of refurbishing the entire building. They'll want to be targeted in their investment."[11] Sorin's recommendation of targeted investments yielding quick returns resonated with the strategies of Old Town's early investors. Mac Popescu, a professor of architecture who founded Club A in his student days, argued that "the basement has clear advantageous for starting out a club: it's sturdy and protected. The basement still has a clear public address, but it doesn't really concern the street, so it doesn't need to look attractive from the street in order to get people to come in."[12] Dumitru, who opened his club underground in 2005, agreed. "Our building wasn't initially safe for commercial activity, but the basement was," Dumitru explained. "By putting the club in the basement, we could ignore the rest of the building for a while."[13] This targeted investment strategy was made possible by the well-established allure of basement bars, drawn as much from Bucharest's own Old Royal Court as from the speakeasies and jazz clubs of New York and London and the cabarets and ratskellers of Berlin.[14] While Nelu, Dumitru, Corneliu, and Mihai regularly played upon the double meaning of underground as both "subterranean" and "countercultural," they were all clear that Old Town's underground roots were not about affecting a sense of style. "Back when we opened in '99, there were only a few clubs in Old Town—Club A, Fire, Backstage—and they were all underground. And it wasn't a cult of the underground," Corneliu insisted. "I mean a space is a space," he continued with the same kind of geometric sensibility that had been expressed by McDonald's corporate managers (see Chapter 1). The underground, for him, provided a blank canvas upon which a variety of atmospheres—countercultural or otherwise—could be generated. "If you play the right music and get the right lighting, you can make anywhere feel 'underground' counterculturally speaking," Corneliu maintained. "The clubs went physically underground because it was good for business."

Beyond the cost of stabilizing and refurbishing Old Town's aboveground buildings, club owners also pointed to the inherited configurations of their historic structures. Simply put, the layouts of ground and upper floors did not conform to the demands of a twenty-first-century nighttime economy. Constructed in the nineteenth century, Old Town's villas originally served as private homes and shops for wealthy merchants and craftsman. The floors of these townhouses were divided into smaller rooms that not only supported the separation of the shop from the home but also prioritized intimate rather than large gatherings.[15] "A quick tour made it clear that when these buildings were built, no one ever thought they would be used for a café, much less a nightclub," Nelu explained. "The space was just too divided, too small." The problem posed by the intimately constructed rooms of old, Nelu and the others attested, was that they put too much pressure on the volume of customers needed to stay solvent. "Most of your money is made on the weekend," observed Mihai, "and you just can't bet on enough people passing through those small spaces above to turn a profit in just two nights."

Old Town's substructures, by contrast, reflected an entirely different approach to the production of space. Merchants and craftsman built their homes upon large cellars intended to store the fruits of old Bucharest's numerous orchards and vineyards.[16] Instead of compartmentalizing the underground into a series of intimate rooms, as they had done with the shop and hosting spaces above, architects produced open floorplans underground, enabling barrels of wine and brandy to be rolled, stacked, and stored. The alternative production of space underground provided twenty-first-century entrepreneurs with an immediate workaround to the limitations posed by the layouts of the floors above. "These large basements can hold two or three hundred people. Down here," Mihai added with a shrug, "you can hold the volume of people you need to make the business work." The others agreed, albeit not without caveats. "You still have to work overtime to do it," Nelu cautioned.

Floorplans, of course, can be remodeled and superfluous walls removed. The organization of space in Old Town's aboveground structures could have (and in most instances have indeed) been completely reimagined, preserving nineteenth-century facades to aestheticize the twenty-first-century configurations installed behind them.[17] These efforts, however, unfolded only after significant investments of money and time. In sharp contrast, Old Town's substructures proved quicker and cheaper to flip. By focusing underground, club owners were able to quickly monetize their buildings while delaying the costly processes of stabilizing and then aestheticizing the ground and upper floors. "Such a large investment upfront creates a lot of liability," Corneliu explained. The neighborhood's substructures, however, had retained their integrity through the neighborhood's decades of neglect. While the walls needed painting, they did not need to be rebuilt wholesale. Old Town's earliest investors realized that they could make quick money by turning toward the underground and make simple renovations to their cavernous cellars. "This cellar was built by the Old Royal court. It's got a lot of exposed brick and ceilings," Corneliu told me as we toured his club at an off-peak hour. "They're solid, stable spaces. They're old but already beautiful in their own way." Nelu shared similarly romantic sentiments while showing me the relatively minimal work he had done to stage his cellar as a club. "The brick cellars from the Old Royal Courts characterize the historic center," he began. "It creates an atmosphere. I just scrubbed the bricks clean and installed the electricity and some plumbing." With just a little lighting, some music, and of course alcohol, Corneliu and Nelu agreed, Old Town's historic brick cellars appeared enchanted. Even a minimal effort at gentrification rendered the grit of an exposed brick cellar into an aesthetic pleasure of the first order. Dumitru, for his part, was less poetic in describing his easy flip. "This isn't a fancy club where people come to be seen," he divulged, drawing a pointed contrast with the high-gloss, higher-cost bars and clubs that have taken shape subsequently aboveground. "I cleaned the place up and otherwise slapped together a bar and some bathrooms. We didn't have much else to invest." Ten years later, Dumitru's club was packed with youthful partygoers in skinny jeans and graphic tees each and every weekend.

Underground Space and Time

As Old Town's nighttime economy began taking shape, club owners noted that their basements and cellars did not merely offer convenient spaces for their businesses. In addition to containing a critical mass of partygoers inside, vaulted cellars also proved exceptional at keeping sound and sunlight out. Disconnected from the cues of circadian rhythms, the underground made possible new, more productive temporalities that unfolded parallel to and distinct from those of the city above. In place of the organic cycles of dawns and dusks that frame the nighttime economy of Old Town's aboveground bars and venues, club owners quickly found that the underground rendered the night's boundaries plastic, malleable, and stretchable in profitable directions. No longer limited by the rising sun, club owners could easily extend the nighttime underground to indulge the hedonistic desires of their most spirited customers. For those spending their weekends partying underground, Saturday night no longer had to end before one wished it to end. The underground's ability to allow consumer desire to conquer planetary rhythms gave Old Town's bourgeoning entertainment economy a novel depth not found elsewhere in the city.[18]

"Being underground changes how you manage the place," Dumitru explained. "It dictates when you're open and when you're not." Each of the other club owners and managers I spoke with agreed on this point. The advantage to operating underground came from the ability of basements and cellars to prolong the critically lucrative window of the weekend, particularly Friday and Saturday nights. "Like any club," Mihai explained, "we draw people every night, but the weekend crowd is something else. Crowds only really come out on Friday and Saturday nights. So, we only really have two nights a week to make a profit." Nelu agreed. "We don't even open during the day," he said. "Obviously people still want to eat and drink when it's light out, but no one wants to spend time in a basement on a sunny afternoon. People will go to a pub to do that." Nelu, like the others, relied entirely on customers coming after dark. "We open at 7 p.m. every night," he continued, "but it's not until after 8 p.m. that people get interested in going downstairs."

The ability of these clubs to turn a profit while conceding the day hinged upon their mastery over the night—their capacity to extend the evening hours in ways that other venues, including Old Town's glossier upscale (and aboveground) bars and clubs, could not. "Being underground lets us stay open until the last customer wants to leave," Dumitru explained. "The venues aboveground begin to close at 4 a.m. and then a wave of customers comes down here to keep their night going." With the usual circadian cues blocked, nights spent undulating down inside Bucharest's basements and cellars easily spilled over into the morning hours. For those not ready to end the party, drawing the evening out underground made intuitive sense. "This is where you come if you haven't already found someone to go home with," one twenty-something woman explained to me in the early morning hours at Club A. The elongated weekend night out underground provided her with a needed release from her public relations job in the city center. Drunk bodies, soft drugs, and the prolonged promise of casual encounters circulated through the sticky floors and sweaty dance halls of these basement clubs. "There's enough energy to keep you going until the Metro starts up again at 5 a.m.," an IT worker in his early twenties explained. He preferred to drink the money that would otherwise be spent on the long cab ride to his new housing block along the city's edge. The mix of young professionals flush with their first salaries, university students, and foreign tourists on a stag weekend bender regularly stretched Saturday nights inside Dumitru, Mihai, and Nelu's clubs to well after seven o'clock on Sunday mornings.

As the depths of Old Town's basements and cellars eclipsed the dawn, holding out the potential of an endless night, their thick walls also absorbed the sounds of revelry so that the party underground did not disturb the morning rhythms unfolding above. The aesthetic repurposing of the underground for the middle classes facilitated a segmentation of the city center that enabled very different people and activities to coexist simultaneously, one atop of the other. While walking through Old Town's narrow, winding streets after dawn, for example, it was not uncommon to see young revelers brushing shoulders not only with joggers and shopkeepers, but also with an older generation on their way to church. Although providing fleeting moments of incidental proximity among different kinds of people, the underground ultimately furthered separation and autonomy. Most notably, Club A is situated immediately next to Sf. Nicolae Șelari church, which offered early morning services. "Some might think it's a contradiction to have a night club next to a place of worship, but not really," Mihai quipped. "The club is underground. It holds in the noise from both directions. The basement keeps our very different services from bothering one another," he said with a smile.

Expanding Upward

The spatial and temporal break with the world above made possible by Old Town's underground basements and cellars helped the emergent nighttime economy establish its roots. The prolonged weekend of Bucharest's emergent club scene enabled early investors to move the requisite number of drinks to make their bank loans and payroll. It also fostered a novel intensity that captivated the capital's new middle classes while also attracting a growing wave of foreign tourists. The clubs' total commodification of the night, from its start to whenever people decided they had finished with it, relied entirely on the space of the underground.[19] "We wouldn't be here without it," Dumitru admitted. And once Old Town's roots took hold beneath the city surface, the nighttime economy began to grow upward, slowly undertaking the costly renovations needed to incorporate the neighborhood's well-worn superstructures and terraces. But rather than extending their successful underground dive bars and clubs upward to the street level, owners kept these parts of their business contained to the basement cellar. Instead, they imagined their buildings' superstructures as separate canvases on which an entirely different atmosphere could be generated for audiences further upmarket. The underground's qualities have proven crucial for practices of segmentation that enable the proliferation of diverse "entertainment products" designed to cater to different segments of the middle classes, often layered one atop the other.

"We looked to open up opportunity in every direction," Dumitru explained as we toured the various levels of his building. As money flowed into Old Town, the early entrepreneurs who had taken hold in Old Town's basements and cellars began stabilizing and repurposing their aboveground floors. Building facades were plastered and painted, walls reinforced, and layouts reconfigured to accommodate larger volumes of people. Just as Club A could be vertically juxtaposed with a church without disrupting either service, entrepreneurs leveraged the underground's insulative qualities to create different strata of experiences within their buildings, all unfolding simultaneously. "Our basement is built so deep, it holds sound incredibly well," Dumitru explained. "We realized [not only that] it keeps us from bothering the neighbors, but also [that] it could keep us from bothering ourselves up above. We could build an entirely different concept above our club and reach an entirely different segment of customer." Dumitru's description of layered growth resonated with other practices across Old Town. "Almost no basement club uses its top floors for a club," Corneliu observed. "What's happened here is that the basement and the ground floors have become their own distinct things."

Corneliu, Dumitru, Nelu, and Mihai approached their properties with an eye toward segmentation in efforts to branch out in ever more profitable directions. Binary segmentations between above and below, and between inside and outside, became the basis for establishing parallel entertainment concepts that could capture the attention of different kinds of people within a single building. "We had a clear sense of what the basement was about: loud music, live concerts, dancing—it's young, it's cheap, it's high energy, and busy at night," Corneliu explained as he showed me through the rest of his building. "When it was time to develop the street," he continued after taking the stairs up to ground level, "we developed a pub concept that would do the opposite. The drinks are more expensive. It's a place to sit and chat with friends and colleagues over a beer or an espresso." The shift in character from basement to street level was partly temporal and partly about the introduction of a new surface aesthetic bringing together oversized mirrors, Edison lightbulbs, and stained wooden bars, tables, and chairs. Rather than selling shots of vodka to send the young into overdrive, street-level extensions played to different fantasies, offering cozy rooms where people could slow down, settle in, and catch up. "With the pub, it's about bringing older clients in during the day," Corneliu noted. Keeping a brisk pace, he walked me through his pub, out the back door, and onto a seasonal garden patio for summer lounging over long drinks. "Each space is specific," Corneliu assured me; "each has its own fingerprint, its own mood." Each space also had its own peak times, opening up to Corneliu the opportunity to capitalize on the day and the night, on those looking to speed up as well as those seeking to wind down. Nelu, Dumitru, and Mihai did the same, segmenting their buildings in order to generate a multiplicity of spaces with different characters and temporalities and to attract different kinds of customers.[20]

The relative success of investors in producing a constellation of juxtaposed, imbricated, and ordered atmospheres in Old Town generated the effect of an overwhelming totality whose collective effervescence resonated with even the most seasoned socialites, such as John, the HBO actor introduced previously.[21] Buildings were packed and streets remained crowded as people found their place. As nights wore on, parties converged in the underground clubs. With Old Town's popularity rapidly spreading abroad, those invested in the nighttime economy declared that their work had only just begun. "All the time I'm looking left and looking right," Mihai explained as he showed me Club A's new aboveground lounge. Customers sipped tall drinks and nestled into deep chairs while instrumental music played softly overhead. "We're all studying the market, building new strategies on top of our base." Old Town's basement and cellars nevertheless served as the social and economic engine of the neighborhood.

Requiem

On October 30, 2015, shortly after 10:30 p.m., a fire broke out at Colectiv (Collective) nightclub, located near Old Town. The club had operated out of a converted factory in central Bucharest without any of the requisite safety inspections. Although not within Old Town itself, the venue resembled that neighborhood's subterranean clubs in a number of ways. Hugely popular, owners regularly allowed the club's audience to swell to its absolute physical capacity—upward of eight hundred people—for live concerts and DJ sets.[22]

On that night, onstage pyrotechnics during a live performance ignited flammable soundproofing that had been illegally installed overhead.[23] The venue's jerry-rigged ventilation system made the fire spread quickly. The ceiling lit up "as if it were a puddle of gasoline," reported one survivor.[24] Flaming debris rained down on concertgoers, who found themselves trapped by illegally blocked exits. "It would have been better to be in hell than there," assessed a firefighter at the scene.[25] Sixty-four people died in the fire and over a hundred more badly burned victims were sent to hospitals that were ill-prepared to treat their injuries.[26] The Romanian media immediately cast the fire as a national tragedy.[27] Public commentary attributed Club Colectiv's ability to operate unlicensed and uninspected to corruption.[28] Suspicions of similar lapses and circumventions of safety protocols instantly engulfed Old Town's basement and cellar clubs.

The tragedy reverberated throughout the circles of architects, engineers, developers, city planners, and center-north office workers in which I moved to carry out this research. With solemn faces and somber tones, many of these young professionals confided to me about their friend or colleague who had been at the ill-fated concert, or of having had plans themselves to attend that night if only they had not gotten caught up at the office. The traumatic scene at Club Colectiv invited middle-class professionals and university students alike not only to identify with the victims but also to wonder aloud if one day they too might find themselves trapped underground and unable to escape the deadly consequences of bribery and corruption. Underlying these speculative moments was the visceral realization of their own semi-disposability: they—as middle-class professionals—were valued and catered to as consumers but not fully protected as rights-bearing members of the polis. The underground clubs marketed toward the young and in the know for their late nights out on the town had proven horrifically prone to dangerous accidents. As one thirty-something architect put it glibly, "I'm sure the Hilton bar is up to fire code," referring to the stodgy hotel bar where he imagined foreign investors gathering when in town. A twenty-something marketing professional who frequented the neighborhood's underground clubs framed her indignity in terms of surface aesthetics. Walking through Old Town shortly after the fire, she remarked to me, "This neighborhood is supposed to represent that Romania has turned a corner, that life here is changing for the better. But now it's clear that this is all a lie. Nothing has changed for us but the coat of paint." Contrary to the sense of national ascendency supported by the regular evenings she had spent out in the capital's now cosmopolitan nighttime district, life after the fire felt as precarious as ever. While many entities happily profited off of her consumerism, they did not appear to be fully invested in assuring her health, safety, and well-being. Others agreed. In the aftermath of the Colectiv fire, tens of thousands took to the streets of Bucharest with signs reading, "Corruption Kills!" (*Corupția Ucide!*), demanding anti-corruption reforms from the government.[29]

Across Old Town, basements and cellars that earlier had invited people to act carefree instead evoked feelings of fear and degradation. Club owners suddenly fielded costly demands to excavate additional exits from their buildings, to install fire doors and sprinkler systems, and to utilize fire-resistant building materials, for example. Owners also could no longer fill their venues to the cellar's physical limits. As the economy of operating underground clubs quickly shifted, club owners wrestled with the realization that their targeted investments with quick returns were in fact far riskier than they had realized. As club owners tried to adjust, they described a landscape in which the government appeared unwilling to accept responsibility for enforcement. "All of these reforms require certification," one owner confided to me with frustration a few months after the tragedy, "and in this moment, City Hall isn't signing off on permits. And so I go to the fire brigade for an inspection, but they aren't signing off on permits. It sounds crazy, but no one knows who is supposed to do the inspecting. It seems I am now the only one responsible for certifying the safety of this property. It makes no sense." Liability rose alongside concrete operating costs. An increasingly skeptical public stayed away from Old Town's underground venues. In the months that followed the fire, many of Old Town's oldest and most storied underground clubs opted to shutter. When possible, entrepreneurs turned their attention to their aboveground operations. "We didn't want the risk of staying open underground," Corneliu explained to me. "We decided to close our basement until we could get our paperwork in order, whenever that may be." Like the other owners with whom I spoke, Corneliu preferred to describe the closure of his basement club as a matter of incompletable paperwork rather than of his cellar's material incompatibility with code. No one admitted, even in hindsight, any risky corners that may have been cut or lapsed judgements. Club A, by contrast, remained open. Not surprisingly, the club originally designed by and for architecture students was able to demonstrate its compliance. "They eventually came back," Mihai told me the following year, his club appropriately effervescent for a Saturday night. Their clients, however, were not as carefree as they once had been. "People have started asking where the exits are, where the extinguishers are kept," he continued. "People are now interested in these details."

Old Town's basement and cellar clubs opened up to Bucharest's growing middle classes a site to let loose in the city center. The development's growing popularity with international tourists even extended to the district a cosmopolitan air of belonging to something that transcended the city itself. As drinks poured late into the night, the excitement of Old Town made it all too easy to avoid asking sober questions about whether the undulating crowd might be jammed in just a little too tightly, the egress stairwells too few in number and too narrow in width, or if the sound system's wiring was too frayed. The Colectiv tragedy exposed the dangers of the compromises to safety and wellbeing that had helped to make underground clubs financially viable sites for the middle classes in the first place. As it became clear how costly it would be to bring these spaces fully up to code—to make them as safe to slosh through as the Hilton bar serving drinks to the well-heeled— many shuttered. Some shrugged at the closing of many of these underground clubs. They had already served their purpose in helping Old Town's new nighttime economy take root. More polished bars, restaurants, and clubs atop the city had become sufficiently established, and the district would continue to expand aboveground. Those who had seen firsthand how generative investments into Old Town's underground had been, however, recognized that its basements and cellars would again be repurposed. "We hope to reopen our basement one day," Corneliu told me at our last meeting. "Underground clubs are the soul of the historic center, even now. The entire neighborhood is built upon basement clubs. It's only a matter of time until they will be brought back to life, in one form or another."

CHAPTER 5

Parking Garages

A longside efforts to move the new middle classes downward, into an aesthetically repurposed constellation of Metro stations, basements, and cellars, the municipality also relied on the underground to solve a different problem tied to the growing middle classes: Bucharest's spiking rate of automobility. As was the case in post-socialist cities from Moscow to Tblisi, the growing ranks of Bucharest's middle classes used their rising salaries to buy into personal automobility, sparking a seven-fold increase in car ownership.[1] For the middle classes themselves, car ownership promised eased mobility, convenience, and status. The municipality, however, understood the rise of middle-class automobility as a liability. When in use, city planners noted, the sheer number of cars physically overwhelmed Bucharest's socialist-era roadway system (see Introduction). Stalled highways and bottlenecked side streets emerged as a perverse sign of economic development.[2] The same cars proved just as problematic when not in use. For much of the twentieth century, socialist-era planners had assumed the predominance of public transit at the expense of personal car ownership.[3] Socialist commitments emphasized not merely public transport but also the collective use of cars in the form of taxis and state-owned rental services.[4] Given that widespread car ownership was then unimaginable, socialist-era planners had not incorporated mass parking into development projects.[5] While the growth of the business services sector in Bucharest brought car ownership within reach of the masses, the development of back offices did not provide the vast majority with the means to live in a newly constructed building with designated parking; those kinds of buildings were marketed instead to highly paid professionals. The impasse between a history of socialist-era planning principles and the rise of middle-class consumerism grew to surreal proportions. In 2009, news

Figure 19. University Square garage, 2017. Photo by author.

accounts estimated that Bucharest had formal parking for only one in five cars; that figure was revised downward to one in ten a decade later.[6] Reports indicated that the parking shortage was most acutely felt in the city center, with an estimated thirty cars for every legal parking space.[7] These figures stand in stark contrast to planning practices in other consumer societies, where ratios of between three and eight parking spaces per car in circulation are common.[8] As the struggle to find formal parking in Bucharest grew nearly impossible, the middle classes resolved to leave their cars wherever they could find room—along side streets, atop lawns, gardens, and parks, or even in the middle of cobblestoned squares—leading in 2016 to an astounding one million illegally parked cars in Bucharest.[9] Officials believed 200,000 of those cars to be either completely abandoned or repurposed as private storage units to offset the constraints of Bucharest's tight housing market.[10]

Among city planners, the rise of middle-class automobility—and the staggering levels of illegal parking that came with it—aggravated interrelated concerns about function, aesthetics, and the continued ability of the city to attract foreign investment. The preponderance of illegal parking contributed not only to Bucharest's traffic problem but also to the diminished quality and character of its city center. As workers, the new middle classes were vital to the city's continued growth, but as they went about their private lives, their automobility threatened to materially overwhelm the city in ways that jeopardized its continued vitality. The municipality expressed concerns that the poor state of the city center might make Bucharest (quite literally) an unattractive place in which to invest. To compete with other European countries for foreign funds, one plan laid out, downtown Bucharest needed, "high quality public spaces which serve the community and strengthen its image as a European metropolis."[11] This plan's admirable appeal to "serv[ing] the community" of course belies fraught and heavily policed questions as to who counts as a member of the community, not to mention how their needs are defined and best served. As demographers projected that the city's parking problems would only worsen in the foreseeable future, city planners converged in 2008 to draft a municipal parking strategy for Bucharest.[12] With an eye on Paris as their model, planning documents laid out the municipality's strategy to "liberate" central Bucharest's congested roads, sidewalks, and main squares from mass illicit parking.[13]

The municipality's parking strategy contained three pillars, the first of which proposed investing some €311 million to develop fifteen underground parking garages beneath the city center.[14] While developing parking underground was undeniably more expensive than aboveground, planners justified the added cost as an investment in the city's aesthetics, and by extension, its financial wellbeing.[15] As one city planner quipped to me, "Bucharest is a European capital—you cannot build a big, ugly car park within eyesight of one of its main squares. [The garage] would ruin it."[16] The municipality also proposed a series of aboveground garages along the urban periphery, where aesthetic concerns were less pronounced.[17] Planners hoped that suburbanites would park their cars in these peripheral garages and navigate the city by Metro, tram, or bus.[18] To encourage drivers to "park-and-ride" on public transit, the second pillar of the strategy encouraged the aggressive ticketing of illegal parking. It also introduced coordinated parking rates across the city that made street parking more expensive than parking in the city center's underground garages, which in turn would be more expensive than parking aboveground along the periphery.[19] Planners hoped that aggressive ticketing and tiered parking rates would encourage the economically rational to move their cars in desired ways—namely, off of city streets. The parking strategy's third and final pillar called for the installation of parking barriers to physically prevent drivers from accessing the city's pedestrian-centered zones, such as Old Town (see Chapter 4).[20] Taken as a whole, the municipality's three-pronged strategy sought to clear mass automobility from view within the city center so that, in place of congested roads and overrun squares, the Parisian fantasy of a walkable city framed by restored nineteenth-century architecture and populated by cafés, bars, and restaurants could take shape.[21] "What counts is that this plan is put into practice," one politician proclaimed, "because for many years the metropolitan area was not how a European Capital should be, not only from the point of view of transit but also of city planning. Development has been chaotic."[22]

The underground provided the needed staging area to remake Bucharest's chaotic character after the ideal European capital: Paris. While much has been written about what Erving Goffman famously called the "backstage" of everyday life, where people prepare for the social roles they must play, Bucharest's underground parking garages bring attention to what the French call the *dessous*: the often overlooked "substage" where the anonymous and unapplauded work of generating atmosphere unfolds.[23] In Bucharest, the gentrification process relied upon the vertical tension of staging the city center by moving middle-class automobility down into a newly constructed substage, one that would allow the middle class's labor and consumerism to be present within the aestheticized city center without their physical presence becoming an overwhelming visible blight upon it.

This chapter describes efforts to convince middle-class drivers to utilize the city center's newly constructed substage: underground parking garages essential for materially staging dreams of a Parisian-styled city center aboveground. Unlike in the case of the aestheticized Metro stations, basement apartments, and subterranean nightclubs already discussed, however, this chapter brings into focus the complicated efforts to expand the urban underground. In so doing, the chapter marks a shift in the book's focus, moving from the process of attracting the middle classes underground to the obstacles that stood in the way of this endeavor. While integral to generating the right kind of atmosphere aboveground, underground parking, this chapter shows, ultimately failed to captivate the imagination of the middle classes. To the contrary, subterranean parking garages posed a substantive constraint upon the fantasy of middle-class automobility without offering any compelling lifestyle or status upgrades in return. These parking garages asked the middle classes to give up the de facto free, illegal parking that kept their cars ready at hand in exchange for paid parking garages of dubious convenience. Gentle strategies of incentivization eventually gave way to the derisive rhetoric of responsibility and entitlement and to concrete efforts at barricading key pedestrian zones to prevent them from being overrun.

Aestheticizing Parking

"We are not here to just add more parking to the city's supply," explained Andrei, a senior manager for Interparking Group, as he adjusted his cufflinks. Founded in Brussels in 1957, Interparking operated 373,455 parking spaces across nine EU countries at the time of our conversation in 2016, generating €437.8 million in annual revenue.[24] Romania represented Interparking's newest market. We discussed Interparking's Bucharest operations over espressos in Andrei's brilliantly white office located two floors beneath University Square. "We are here to introduce a 'European parking experience' to Bucharest," he explained. As became clear over the course of our conversations, "European experience," for Andrei, referred as much to an upgrade in the material quality of Bucharest's parking spaces as to a shift in the citizenry's parking practices to better align with European norms.[25] Interparking's management had assumed that the introduction of their state-of-the-art parking garages would persuade Bucharest's drivers to move their illegally parked cars into paid parking underground. Their garages' aesthetic appeal to middle-class cultural experience, however, failed to incentivize the city's drivers to undertake the thankless substage work of clearing the city's squares, lawns, and sidewalks of their cars. Underground parking proved too difficult to sell as a marker of middle-class respectability, despite Interparking's best efforts. And as appeals to respectability politics ran flat, calls for more diligent policing soon followed.

"As it stands," Andrei continued, "when most Romanians park, they do not respect the sidewalk; they do not respect the landscaping; they do not respect the flowers; they do not respect the city. They just park wherever they want," he chided. Andrei's assessment of illegal parking cynically elided the infrastructural constraints that confronted Bucharest's drivers: namely, the staggering absence of general parking stemming from Romania's engagement with socialism. It also glossed over the degree of wealth and status required to buy into the designated parking then incorporated into upmarket office and residential developments. Instead, Andrei placed responsibility for illegal parking squarely onto individual behavior, concluding that the masses of middle-class drivers had been making the vulgar decision to park where they wished rather than where they ought. Andrei presented Interparking's "European parking experience" as the needed correction to Bucharest's parking problem, which encompassed both the city's inadequate supply and the apparent déclassé inclinations of its drivers. For Andrei, Interparking garages would enable more of the middle classes to utilize designated parking like their counterparts in Paris, Rome, Berlin, and Brussels, for example.

On November 15, 2012, Interparking opened the first of the city's proposed underground parking garages in central Bucharest beneath University Square. Extending three levels downward, the garage added four hundred twenty-five car spaces and twenty-one motorcycle spaces to the city's parking supply at a total cost of €10 million.[26] Interparking heralded the opening of its University Square garage as restoring order and beauty to the streets of Bucharest. As their annual report boasted, with all of the restraint of a paid magazine advertisement, "Safe and comfortable parking in the vibrant centre of Bucharest? Thanks to Interparking, it is now possible. . . . Thanks to this underground car park, University Square—which previously was always jammed with parked cars—has been restored to a spacious and inviting town square."[27] In 2016, Interparking further expanded its footprint with a second garage near University Square, located beneath the Intercontinental Hotel and the National Theater of Bucharest. Originally constructed in 1969, Interparking invested €3 million to renovate the existing two-story garage. While in no way meeting the city's parking demand, the renovation nevertheless added an additional 969 spaces to the city center's parking supply.[28]

Interparking's investments provided much-needed additions to the quantity of parking spaces in central Bucharest. However, in its promotional materials and interviews with local media, as well as in my own conversations with Andrei, Interparking framed its intervention into Bucharest's parking problems in terms of improving the material quality (rather than the quantity) of the city's parking.[29] As we toured Interparking's University Square garage in 2016, Andrei looped around time and again to the garage's clean design and technological sophistication. "'CO$_2$ neutral' certified, energy-efficient lighting, charging points for electric vehicles," he boasted. Over the course of the tour, Andrei did his best to couch personal automobility and its accoutrements in terms of middle-class commitments to environmentalism and corporate responsibility. "These are the standards of our car parks in France, Belgium, Germany, Austria . . ."—his voice trailed off to inject a pregnant pause. "And now we've brought that parking experience here, to Bucharest." Outside observers affirmed Andrei's talking points. Shortly after opening, Interparking received a European Standard Parking Award for its University Square garage.[30] Beyond bringing the "premium" parking experience of Paris and Brussels to Bucharest, Andrei emphasized that Interparking's operations in Romania charged a fraction of the price. "It costs less than three lei [€0.50] an hour to park here while an espresso in Old Town costs around ten lei [€2]," Andrei continued, drawing an unintuitive parallel between parking and café culture as recognizable status markers. "So, we are providing the first-class parking experience found in Paris, Brussels, or Rome to central Bucharest for less than half the cost of an espresso," he said with a smile, adding, "We provide an incredible service."[31]

Unlike the chic cafés overlooking the boulevard where one might sip a cappuccino, however, Interparking's underground parking garages never registered as a place to see and be seen. City planners and Interparking representatives alike expected that the promise of a first-class parking experience would move the middle classes to undertake the work of the substage, generating a Parisian atmosphere by clearing illegally parked cars off of the city's surface and out of sight, into paid parking beneath it. Convincing the middle classes to undertake this work, however, proved to be more difficult than municipal and market actors had initially expected. Despite the overwhelming need for parking in central Bucharest, premium or otherwise, Interparking's University Square garages remained largely vacant. Three months after the University Square garage opened, one editorial reported with astonishment that only two hundred of its four hundred and twenty-five spaces were occupied on a Friday afternoon.[32] Assuming that empty garages meant a lack of public awareness, the municipality installed additional signage indicating that sidewalk parking was illegal. Editorialists, for their part, lampooned those efforts, insisting cuttingly that "drivers should already know better."[33] Through 2019, the University Square garages remained persistently half-empty throughout the workweek and even more so on weekends; side streets, by contrast, remained double if not triple parked. Newly opened garages along the urban periphery fared even worse. In 2018, the Metrorex corporation invested nearly €77 million to develop a three-story aboveground garage at the city's northwest edge, next to the Străuleşti Metro stop. Officials hoped that the garage would encourage suburban drivers to ride the

Metro into the city and further that it would provide a place for city residents to store their car away from the center. Two months after its opening, however, the garage operated at under 10 percent capacity while worried officials announced plans to cut the garage's one leu (€0.20) hourly rate by half.[34]

Given the tremendous cost of developing underground garages, their persistent emptiness sparked widespread criticism. Scholars voiced skepticism about steep investments of public funds to build underutilized and unwanted infrastructure while opinion writers turned their ire onto the perceived poor comportment of the city's drivers.[35] "Many drivers just prefer to leave their cars on the street instead of paying for a civilized place in an underground car park," one columnist glibly stated. The rest of the editorial doubled down on the language of (in)civility, framing underground garages as "civilized alternatives to climbing cars up onto curbs and other acrobatics."[36] The introduction of a parking experience on par with Paris or Brussels did little to align middle-class parking practices with European norms. Sidewalks and squares remained jammed with illegally parked cars while spots in underground garages sat vacant. Bucharest's problems of mobility, stasis, and aesthetics continued unchanged. Not surprisingly, municipal and market actors turned to more forceful tactics after appeals to respectability politics fell flat. "Once people start using the underground parking garages, Bucharest will be more like other EU capital cities—like Paris, Brussels, or Rome. The sidewalks will be open and accessible, and, in a way, Bucharest will become more civilized," Andrei insisted. "But to achieve that vision, it will be up to the police to enforce order."

Policing

While the truly elite moved between new developments with integrated parking, the middle classes were left on the move but with nowhere to park. Amidst the overwhelming amount of illegal parking that inevitably followed, the municipality formed a new police outfit, the Local Police (Poliția Locală a Municipiului București), to ticket the behavior.[37] "The University garage isn't filling up the way City Hall would have liked," explained Cezar, a chief for the Local Police. We sat in Cezar's office in central Bucharest in the winter of 2016. "And so we have been given the thankless duty of ticketing those who park illegally." The open resignation in Cezar's tone reflected the fundamental absurdity underlying the Local Police's daily responsibility to curb illegal parking in a city wholly dependent upon it. Newly opened parking garages, important as the added supply of parking might be, delivered nowhere close to the number of parking spaces needed to address the city's total parking demand. "All those cars from the suburbs join all the cars already in the city and there just aren't enough spots. The vast majority of drivers in this city trying to get to work or to the store are forced to park illegally and risk being fined," Cezar conceded. Though he was tasked with enforcing parking law, my conversations with Cezar were nevertheless framed with a sympathy for the plight of the middle classes; while not mad, he was still disappointed in them. Like Andrei, Cezar largely posed the problem of illicit parking in terms of personal responsibility and middle-class entitlement. "Unfortunately," he continued, "the lack of parking has given drivers the impression that they have the right to leave their car wherever they like." Even after the opening of the University garage, illegal parking remained a regular feature of automobility in Bucharest. And although those with integrated parking at their disposal called upon the Local Police to crack down on the rest of drivers, Cezar in no way expected his force to curb illegal parking practices entirely. "People come to me and demand that I take action to get drivers to use the garages," he continued. "And we do write fines. Lots of them in fact. But we are not gods. We can't collect the cars on the sidewalks in our arms and move them ourselves. Even if we could," Cezar added, "we still have nowhere to put them all."

While noisy commentators and disappointed investors made biting remarks about the civility of drivers who parked illegally as boutique underground garages remained empty, those charged with holding drivers accountable still showed some awareness of the impossible pressures bearing down upon the middle classes. Overworked and underpaid compared with their counterparts in Paris and Brussels, the added cost of manning the substage appeared too much to bear, and with too little compensation in return. "And why would drivers park underground?" Cezar asked rhetorically. "In a society like ours, everyone is in a hurry. They're trying to get to work on time, trying to get home, and somehow trying to squeeze a trip to the store in between," he continued, no doubt giving voice to his own daily tensions and frustrations. "People are trying to move as quickly as possible to make their days work, and so they are parking as close as they can to where they're going. But parking garages make people pay, and they make people go out of their way, and they aren't even big enough," Cezar admitted, recognizing how the middle classes hustled to make their days work. In a city where illicit parking was not an isolated problem but rather endemic to the city's operation, buying into legal parking, however green and technologically forward the experience may have been, appeared even to the police as a mismanagement of personal time and resources. And so Cezar policed middle-class parking with a velvet glove. "Realistically, we can't stop illegal parking. We don't even want to make money off of fines," he insisted. "Our goal is to make drivers aware that how they park their cars can turn into a social threat that harms others: illegal parking can kill pedestrians and hold up drivers. Those are the problems we want to prevent."

With garages half empty, and the profitability of such investments drawn into question, the planned construction of the additional underground parking garages beneath central Bucharest stalled. "It is a vicious cycle," assessed Luca, a developer contracted to build three of the proposed underground garages. "People are parking illegally because there isn't really enough legal parking, and there isn't enough legal parking because the existing garages haven't turned a profit. Since it's not profitable, businesses say the state should offer massive subsidies to make more garages, and the state says that park-

Figure 20. Barricaded sidewalk, 2016. Photo by author.

ing personal cars is a private responsibility. And so nothing is moving forward." As questions of personal responsibility and middle-class respectability proliferated and the growth in new parking spaces faltered, Bucharest's sidewalks remained overrun and its roadways bottlenecked. General opinion converged on a basic point summarized neatly by a city planner named Sorin: "This isn't how a European capital should be. People keep buying cars in a city that was planned without any personal parking. It makes for a very heavy heritage from the communist era."

Unable to sell the middle classes on the fantasy of a premium underground parking experience, or to drive them underground with the soft stick of ticketing, the municipality became increasingly blunt in their attempts to curb illegal parking. The municipality installed concrete pillars at about knee-height along the curb of the Old Town development and other pedestrian-oriented zones in the city center. The effort physically prohibited drivers from parking in specific historic squares as part of a broader effort at preserving and promoting Bucharest's eclectic architectural heritage (see Chapter 6). The Local Police also layered onto barricaded streets and squares video surveillance technology capable of recording license plate numbers and mailing fines to a car's registered address.[38] This turn toward targeted barricading of the city center signaled the municipality's abandonment of a holistic approach to clearing the city. Instead, they ceded some ground for middle-class street parking while focusing efforts on narrowly delineated areas. Cezar noted, for example, that the Local Police typically looked the other way at illegal street parking so long as drivers leave a meter of sidewalk for pedestrians.

While targeted enforcement has been successful at keeping large volumes of cars off of prestige sites such as University Square, Cezar nevertheless marveled at drivers' tactical workarounds. It is not uncommon, he explained, for residents to undertake the extraordinary tactic of digging up these cement pillars in order to transform a patch of sidewalk into an overnight parking spot for themselves. Once I became aware of this practice, the evidence of the tremendous effort undertaken to dislodge and displace such pillars became readily visible to me along the quieter squares and side streets of central Bucharest. The arduous act of digging and dislodging appeared as a rejection of respectability politics. "Those are very easy tickets for us to write," Cezar quipped. "But the simple fact," he concluded, "is that those from Bucharest know the zones where they live and work. They know where they can park and where they cannot."

Disembedding University Square

Given the difficulties in curbing middle-class dependency on illegal parking, the municipality used barricades to "disembed" specific pedestrian zones from the illegal parking prevalent around them.[39] Disembedding allowed the beneficial work of the substage to be targeted upon key central squares, most notably University Square. The combination of ticketing, barricading, and CCTV monitoring sufficiently cajoled drivers to undertake the work of clearing their cars off of University Square and moving them underground—or at least onto a side street where police would turn a blind eye. To the extent that the middle class's automobility has been excised to the substage, their more capital-friendly presence as consumers has been monetized in ways that burnish the city's investment-friendly image. To that end, the municipality issued an international competition to renew University Square as a prestigious public space. The winning design, put forward by a Romanian architect and two colleagues who touted their experience working in Paris, rearranged the square as an exclusively pedestrian island with a market at its center. The clean arrangement, commentators noted, placed a heightened importance upon four statues that had long stood across the square's north-

Figure 21. Liberated from illegal parking, 2017. Photo by author.

ern border in an effort to foreground a history of place.[40] "We wanted a simple and elegant solution that highlights the architectural spirit of the Square," explained the Romanian architect Simona Dirvariu to the press. "We wanted to offer a setting that the inhabitants of the city can animate and beautify through their presence"—their presence, notably, as consumer-pedestrians rather than as drivers.[41] The jury praised the architects for their proposal's simplicity, offering an arrangement that celebrated the square's nineteenth-century heritage without being subordinated to it.[42]

In the years that followed, the renewed University Square has reemerged from having been a de facto parking lot to become once again an everyday public space. It has also become a popular stage for larger public events. In the winter, for example, University Square has hosted a Christmas Market. Warmly lit by the flames from gas space heaters, visitors are invited to fill up on gourmet burgers and chocolate filled crepes (*clătită*) or sip mulled wine (*vin fiert*) while meandering around the square's towering Christmas tree. Radio ZU, a local pop music station, broadcasts live from the market and features live performances on a temporary stage. In the quieter months of spring and summer, cafés along the square's southern border spill out onto the square, providing shaded tables and chairs where customers can sip cold drinks and see and be seen.

True to form, the substage provided some of the critical infrastructure necessary to clear the square of the illicit parking common elsewhere in the city center. The substage has helped to set the desired scene, including helping to generate just the right kind of atmosphere on the main stage just above it. Rather than the earlier disorder and dysfunction, the renewed University Square resonated with a carefully crafted image of old-Europe charm punctuated by cafés and Christmas markets, and framed by an eclectic mix of nineteenth-century architecture.[43] While illegal parking still persisted just around the corner, University Square itself successfully emerged as the very kind of attractive public space city planners called for: one that resonated some of the time with the aesthetic and commercial needs of some city residents, but always with those of foreign investors.[44]

CHAPTER 6

≡

Ruins

I n the fall of 2010, construction workers had only just begun to dig the foundation of what would become the University Square parking garage when the first skeleton appeared. Another skeleton was then followed by another and still yet another.[1] The discovery of a cache of human remains beneath the square brought the whirlwind pace of construction to a grinding halt. The industrial excavation of the parking garage's foundation gave way to a slow and precise archaeological investigation, descending painstakingly, layer by archaeological layer, month after month, with the hope of gaining some insight into the origin of the remains discovered within the underground's stratigraphy. By the time archaeologists finished their salvage excavations in 2011, they had uncovered some 624 skeletons—each of which had been carefully arranged for its individual burial. Lab analysis dated the remains to between the sixteenth and the nineteenth centuries. Their placement, read against historical documents, suggested that the construction crew had stumbled upon the cemetery of the Sf. Sava Monastery.[2] The cemetery's discovery excited archaeologists, who hoped to find in the unearthed skeletons new insight into Bucharest's medieval origins.[3] They speculated that additional material evidence of historical and scientific significance might be discovered as the excavation of University Square progressed, such as Sf. Sava's building foundations or perhaps some partially preserved walls of Romania's first institute of higher education, the Royal Academy.[4] As archaeologists enthused about the historical and scientific import of the existing finds, City Hall's Director of the Department of Investment publicly announced that the development of the parking garage would soon recommence, its plans unchanged. "We are at the stage of uncovering the last hundred meters of

Figure 22. University Square excavation, 2011. Courtesy Wikimedia Commons.

space for the construction of the parking garage," the director explained to reporters. "The new discoveries will not lead to the modification of the parking garage project at University Square," he assured them.[5] The archaeological finds would pose no further delay to the garage's construction.

In their effort to curate an idealized image of old Europe in central Bucharest—one that would capture the attention of foreign investors—municipal and city planners worked to preserve and enhance the city's architectural heritage (see Chapter 5). Aboveground historical remains were an obvious asset. "The heterogeneous and eclectic character of the center of Bucharest must be emphasized in order to create an urban identity that residents are proud of and which will attract tourists and investors alike," reads one development plan for Bucharest's historic center. "With high quality public spaces [and] a very diverse architectural heritage . . . the center of Bucharest will be an identity mark of a European metropolis with a special history and an unexplored potential."[6] Planners sought to "liberate" the city center's architectural splendors from the excesses of middle-class automobility by way of the underground, which they saw as an available repository that could absorb the illegally parked cars in and around University Square. However, the construction of the University Square garage laid bare the simple fact that the underground is not empty. For one, it is filled with rock and soil that must be painstakingly removed, but it also contains archaeological remains whose protections under Romanian and EU law threaten to bring the pace of foundation laying to a full stop.[7] As archaeologists took control of the site, they slowly and deliberately cleared stratigraphic layers from across a trench, with an eye to understanding how the city came to be. Each stratum received careful documentation in notebooks, including textual descriptions, illustrations, and photographs.[8] While recent technological advances in stratigraphic excavation have allowed for greater insight, these innovations have only further slowed the pace of archaeological clearance by enabling additional subdivision of the stratigraphy into micro-stratigraphy.[9] Alarmingly for developers, the methodical work involved in assessing the historical value of archaeological remains delays construction timelines and pushes projects over budget. It even raises the possibility that multimillion-euro investments might be permanently stopped mid-construction. Such tensions point to a fundamental misalignment between the work of archaeology and the work of development that is specifically temporal in nature: whereas developers seek to clear out the underground quickly and with an eye to the future, they must contend with archaeologists who turn to the underground with an eye to the past, and at a comparatively glacial pace that poses a threat to development plans.[10]

Focusing on University Square, this chapter brings into focus the temporal obstacles that impede the clearing and developing of Bucharest's underground, from parking garages to new Metro lines as well as to the foundations and basement levels of office and residential towers. It takes as its point of departure Laura Bear's argument about time as a critical technique of capitalism, observing how elites produce space and accumulate value by exerting control over the pace and rhythm of production.[11] This chapter argues that producing the underground as an empty space that can be profitably gentrified and harnessed toward the city's vertical ordering requires overcoming the deep and deliberate temporalities of legally mandated, systematic archaeological excavation. Building upon the previous chapter's analysis of the barriers faced by efforts to attract the middle classes underground, the pages that follow bring into focus governmental and corporate strategies aimed at speeding up the slow temporalities that characterize the creation of underground spaces in order to match the frenetic clip of capitalist development and the middle-class lifestyle those spaces produce. The gentrification of the underground, this chapter shows, is an effort not only to control where people do things but also when and how quickly.

Discovery

The uncovering of human remains from the Sf. Sava cemetery beneath University Square quickly devolved into a circus. "Someone had called the police and reported the excavation," recalled Alexandra, a senior researcher at the Romanian Academy's Institute of Archaeology (Institutul de Arheologie "Vasile Pârvan"), when we spoke in her office in 2019. Alexandra had helped to oversee the excavation at University Square. Although archaeologists had installed privacy paneling around the site's perimeter, a curious passerby was still able to peer through and glimpse exposed human remains. Convinced a murder had taken place, the concerned citizen called the police. "I had an officer telling me to stop what I was doing, that I was disturbing a crime scene," Alexandra chortled. She rolled her eyes over the lack of coordination between government officials. While archaeologists conferred with the police about the details of the site before them, television crews started to arrive. "They broadcasted live next to the square for an entire day," Alexandra continued. Sensationalized accounts followed in the news with reports of a discovery "as macabre as it is spectacular" that would shed light on Bucharest's "forgotten history."[12] Adopting a knowing tone as she stirred her tea, Alexandra added with a shrug, "Human remains always trigger the imagination."

For those involved in the excavation of University Square, however, the discovery of the Sf. Sava Monastery and, later, of the Royal Academy were of no surprise at all. "There were plenty of clues as to what might be down there before construction got started," Alexandra continued. Recent excavations to install telecommunication cables had already unearthed parts of the Sf. Sava cemetery and of its nearby church. In another conversation a few months later, a senior researcher at the National Heritage Institute (Institutul Național al Patrimoniului) named Elisabeta explained to me that even relatively shallow construction in the historic center regularly turns up troves of remnants from the seventeenth to the nineteenth centuries. "I helped to oversee the canalization of French Street [Strada Franceză]," Elisabeta began, referencing a nearby location in Old Town. "We found a great number of foundations, wood pillions, old streets, and the caves and tunnels that merchants dug to store their wares, and we can connect these finds to the folklore about historic

Bucharest," she continued.[13] Elisabeta spoke enthusiastically about these finds, as though their historical significance was self-evident. Those invested in redeveloping historic Bucharest toward a brighter future shared Elisabeta's expectations if not her excitement about uncovering chance finds. "It's an old town," stated Luca, whose construction firm has engineered some of the deepest foundations in Bucharest (see Chapters 5 and 7). "In the center developers constantly come across artifacts. And that's a big problem," Luca offered with a heavy sigh, "because the law requires that we stop construction so that archaeologists can do their studies."

The ruins of Sf. Sava, heritage experts argued, are of particular significance to Romania's national narrative. At the time of their discovery, members of the Institute of Archaeology recommended that the construction of the parking garage be stopped entirely.[14] "Sf. Sava was important because it could be connected to great personalities and a great moment in Romanian history—to the creation of the first Romanian university," Alexandra explained to me. "The ruins were connected to the statues of Gheorghe Lazăr, Mihai Viteazu, Spiru Haret, and Heliade Rădulescu, which stand proudly in the square. And so there was a good amount of excitement because they affirmed the idea that we are an old city, that we have a deep tradition." In addition to signaling Romania's historic commitment and contributions to higher education, the site evidenced, according to archeologists, the economic importance of medieval Bucharest as a trading hub for the Habsburg and Ottoman Empires as well as the Kingdoms of Poland and Venice.[15] Nevertheless, neither municipal bureaucrats nor the garage's investors planned to alter the garage's construction. Their commitment was oriented instead toward making Bucharest a future trading hub within the European Union and the broader global economy. Amidst a public back and forth between preservationists and developers, the city's editorialists could only wonder glibly if the space beneath University Square would become a parking garage or if it would indefinitely remain an archaeological site.[16]

Bucharest's recent history gave archaeologists some precedents for preserving medieval ruins in the face of developmental ambitions. The fifteenth-century ruins of the Old Princely Court, located just down the road from the Sf. Sava discovery, faced near-certain destruction in the 1980s amidst Ceaușescu-era efforts to rebuild Bucharest as a monument to Romanian socialism.[17] Archaeologists, however, successfully preserved the ruins by appealing to their importance in supporting the national narrative desired by the Romanian Communist Party. Bucharest's sprawling formation at the convergence of several mercantile and artisanal camps had troubled Romania's socialist-era leaders, who sought to build a socialist future upon a centralized political and economic system.[18] As Ema Grama details, even though Bucharest had never had a clear city center, archaeologists successfully preserved the Old Court by claiming that it functioned as the historic nucleus of Bucharest: "its place of origin, its axis mundi."[19] Recast as the city's historic center of politics, Ceaușescu-era officials became convinced that the Old Court was worth preserving insofar as it provided them with a material past upon which to ground their project of building a socialist future. "Ceaușescu was not interested in the history of the medieval city as such," Elisabeta assured me over coffee. Her opinions were shared by other historians and archaeologists with whom I spoke. "The medieval city, the Old Court, were simply useful to him. Otherwise, if he actually cared about the past, he would not have demolished so many historic neighborhoods and churches and monasteries," she insisted.[20] Despite Ceaușescu-era efforts to remake central Bucharest in a different style and scale, officials preserved the Old Court as a material testament to the city's past and its projected socialist future.

As the excavation of University Square unfolded, archaeologists from the Institute of Archaeology similarly pointed to the historical importance of Sf. Sava and voiced the opinion that the University Square parking garage either should not be built or should be redesigned to preserve and showcase the uncovered ruins.[21] "It would not have been too difficult," Elisabeta opined after the fact. "The ruins could have been kept and incorporated into the garage either in full or in part. There are plenty of examples of this being done not far from here," she explained with reference to Athens. The construction of the Athens Metro entailed significant archaeological excavations, and officials absorbed costly delays to realign routes and to incorporate finds into its Metro stations, transforming active transit hubs into archaeological galleries.[22] Similar efforts have been successfully undertaken in New York and Moscow.[23] The extraordinary volume and significance of finds at the site of the Yenikapi Metro stations in Istanbul led officials to upgrade their initial plans for a "station-museum" into a standalone multistory museum.[24] After much delay and some debate, however, University Square's parking garage was ultimately built without significant alteration to the original plan. The then mayor "insisted a great deal on pushing the project forward. He went through all the advisory commissions, some research was done, and then they built the garage more or less as planned. The ruins practically disappeared," Elisabeta summarized. Adopting a tone of disappointment, she chided,

"When you have uncovered some structures that give you an idea of the way people lived two hundred, three hundred, four hundred years ago, it seems evident to me that they have value and that it is very, very important that they must be preserved." Elisabeta's colleague at the National Heritage Institute and a professor at the university, Dinu, voiced nothing short of outrage over the decision. "I told my students that to destroy the archaeological heritage of Sf. Sava and of the Royal Academy—which are contemporaneous with the founding of another great city, New York—is nothing short of a crime!"

The Value of Heritage

In the several decades between the preservation of the Old Court and the destruction of the Sf. Sava ruins, government officials had come to assess the value of heritage in a markedly different way. At the end of a socialist modernity, municipal leaders valued and preserved the Old Court for its political utility in crafting a desired national narrative. Following Romania's transition toward a globally competitive marketplace, however, officials acted to build over the ruins of the Sf. Sava Monastery. Unlike the city's architectural heritage, the remnants of Sf. Sava did not contribute to the imperative of economic growth in any intuitive way. Over and above cultural and political considerations, the economic contribution of national patrimony had become its predominant value, determining what is protected and promoted and what is not.[25] Municipal planning documents advocate protecting and preserving the city's architectural heritage for its ability to "strengthen the value of the eclectic character of central Bucharest," one that is believed "to attract tourism and investors, to create a positive brand, to develop a sense of community and belonging."[26] Harnessed correctly, City Hall hoped Bucharest's architectural heritage would serve as a developmental "gem," "engine," and "catalyst," transforming the urban landscape itself into an "economic vector" capable of generating development.[27]

As economic value became the predominant framework for assessing national patrimony, governmental and market actors demonstrated an interest in protecting and preserving architectural heritage at the expense of archaeological remains, despite the latter's cultural and educational value. "It's just easier to explain to a developer as well as to the general public why you would have to preserve something that can be seen—like an old building—rather than something that isn't so easily visible," Alexandra offered. Buried under-

ground, archaeology mattered less in part because its finds were largely invisible and thus less easily leveraged as a source of economic value. "People
can see an old building, relate to it, and understand its beauty," she continued, "but if the discovery is paleolithic and it's a collection of pottery shards—
even if it's a significant find—people will say, 'oh, my grandma has pots.'"
Elisabeta at the National Heritage Institute agreed. "There's no such thing as
preservation archaeology in Romania," she insisted. She struggled to identify a single discovery beyond the Old Court that had prompted an alteration to development plans. "City Hall might stop a project if a medieval wall
made of gold were discovered," she pressed sarcastically, "but even then, it
would only be because the wall was made of gold." With a shrug, Elisabeta
insisted that archaeological evidence and the insights it offers into the past
were simply not important to investors. As a result, planners and developers
worked quickly to clear the underground, minimizing the threat of delay.

The evaluation of heritage through an economic framework cast architectural heritage as an asset to be leveraged and archaeological heritage as
an obstacle to be overcome. "We try to know what is protected from the start,
at the very moment we consider buying a plot of land. We want to see what
must be preserved," explained Boian, who owned a large development company in Bucharest. Understanding a potential site's proximity to protected
buildings was a part of his company's purchasing practices. The assumption
that development must work around the city's architectural heritage had not
always existed in Bucharest. During the socialist era, central planners interpreted many of the city's Orthodox churches as negative heritage. Teams of
architects and engineers actively moved churches hundreds of yards back
from main boulevards and squares to create room for new constructions
deemed pertinent as much to the project of hiding Bucharest's religious heritage as to the advancement of socialism.[28] "But if you're developing an old
site, like in Central Bucharest, you never really know what you will find belowground until you get started. That places a question mark over the entire
project," Boian explained in reference to the legal protections afforded to uncovered archaeological heritage. "You just have to dig and wait to see what
you find. You have to take into consideration the unknown," he concluded
with a hint of trepidation. The unearthing of chance finds slows progress,
throws off timelines, and blows up budgets.

While chance finds present developers with an unknown, developers have
grown confident that such discoveries will not derail their investment entirely. "Once construction uncovers a find," Elisabeta explained, "we [the

archaeologists] are brought in. We make an archive of 3D scans. We recover everything that can be transported to a lab for analysis. And then developers do what they had proposed to do from the beginning. The fact is that never—not even when impressive structures like Sf. Sava emerge from the excavation—never does City Hall pose the problem to developers of preserving and restoring them." Alexandra agreed. From her perspective, municipal and market actors treat the slowed time of archaeological analysis as an obstacle to be overcome by the fast pace of development. "Underground heritage is not valued at all," Alexandra flatly explained. "It triggers certain requirements to be fulfilled. And then we get blamed when [construction] projects miss their deadlines and go over budget." After a pause for effect, she added, "And then we get accused of standing in the way of progress."

An empty underground available for development is, ultimately, a production of gentrification rather than its convenient point of departure. The timely emptying out of legally protected subterranean heritage unfolds through a process that favors development. "The word of an archaeologist isn't taken so seriously," Elisabeta insisted. "I could be called out to investigate old city walls that have been uncovered. It doesn't matter how I speak about those walls. Everybody can see that they are old, that they are significant. And yet they'll still be disappeared to make way for a parking garage." Unlike the eighteenth-century square that must be protected from parking to craft a certain image of a medieval European city, Bucharest's actual medieval walls, foundations, and cemeteries are regularly cleared to make way for the city's present and imagined future needs. The consistent disregard for the heritage underground left Elisabeta shaking her head.

Producing Emptiness

"Archaeology in Romania has become increasingly privatized," Alexandra explained. I had asked why archaeological heritage never seemed to impact development despite being regularly in the way of it. As the rate of development escalated in Bucharest, both Alexandra and Elisabeta attested, archaeological practices shifted in Romania to quicken the pace of analysis and to minimize the risks to development posed by archaeological finds. In 2003, for example, the number of privately funded "preventive excavations" in Romania achieved parity for the first time with scientifically funded "systematic excavations." By 2005, the number of preventive excavations had nearly doubled while the number of systematic excavations had declined.[29] The growing predominance of preventive rescue archaeology over systematic research archaeology reflects the escalating number of development projects requiring archaeological clearance. Archaeologists affiliated with academic institutions or museums typically oversee these excavations. "We're happy to participate in these kinds of projects," Elisabeta offered, "because they provide the opportunity to dig. We wouldn't otherwise have the funds to make these excavations scientifically. And of course we're principled about the work because not only our own reputations are at stake but also our institutions'." The sheer number of investigations raised by Bucharest's construction boom, however, has outpaced the capacity of affiliated scholars, leading to the formation of a busy market for unaffiliated freelance archaeologists. By all accounts, the increasingly privatized structure of archaeology has shifted the practice of excavation and identification of heritage. "The concern," Alexandra began, "is that private archaeology diminishes the quality of the work completed by introducing financial incentives to finish investigations quickly. Obviously, a willingness to make compromises gets rewarded with more contracts."[30]

Alexandra, Elisabeta, and others spoke of increasingly disparate practices guiding rescue archaeology and scientific excavations. "We now have two lives," Alexandra attested: "one is this academic life, and then we have this other persona . . ." Alexandra's voice drifted off as she waded into sensitive waters. "Unfortunately, the standards between the two types of activity are very different." Rescue archaeology is fundamentally constrained in its purview, Alexandra pointed out. Whereas permits for scientific studies empower archaeologists to excavate large areas, private archaeology substantively narrows the scope of their observation. "You're limited to observing what the developer is doing," she explained. "You cannot dig outside of the construction plan. So the work is limited to saying, 'well, you found something and so you need to stop,' which no one is excited to hear." When discoveries are made, there is pressure to complete assessments quickly, leading to aggressive uncovering practices that negatively impact the site. "For example, no one would dream of using a bulldozer when doing systematic archaeology," Alexandra explained, "not even for uncovering the surface. But when you're doing 'rescue archaeology,' if you don't use a bulldozer you're dead in the market because you can't meet any deadlines." With an air of pragmatism, she added plainly, "And then good luck winning the next contract."

Accelerated digging on compressed schedules, practitioners maintained, reduces the potential for uncovering the kinds of significant finds that delay projects and throw budgets out of balance. At the time of these conversations, the municipality was completing construction on the M5 Metro line, which runs from southwest Bucharest through the center and toward the eastern border of the city. Although the M5 cuts through historic portions of the city, the construction had not officially made a single discovery that would pause its progression. "There was absolutely no archaeological evidence along the M5 route," one engineer familiar with the project remarked to me with a categorical certainty. Alexandra questioned the claim when I reported it back to her. "I think it would be more accurate to say nothing has been recognized as such," she said with a smirk. "At the depth they're digging, in the parts of the city where they're digging, my guess is that they encountered paleolithic or perhaps some sort of prehistory that they wouldn't have recognized. If an archaeologist isn't always present, if the [construction] crew is working quickly, if they're digging at night to stay on their timeline, then sure—nobody's going to 'find' anything because nothing was recognized." In this way, development pulls the slow temporality of excavating the underground into the accelerated pace of capitalist development.

When potentially significant discoveries are uncovered by chance, Romanian law requires that construction permits be temporarily suspended while archaeological researchers complete an analysis.[31] "The process starts from the Ministry of Culture (Ministerul Culturii)," Alexandra explained. "It's up to them to recommend what is to be done." Although the Ministry of Culture is formally in control of the evaluation process, the process itself, heritage practitioners noted, unfolds in ways that favor development. Most concretely, Alexandra pointed to a shift in the qualifications of those placed in decision-making roles at the Ministry, which, she maintained, is increasingly populated by political appointees rather than experts. "Our ministers have gotten less and less educated over the years," she remarked. "If in the 1980s and 1990s personalities of Romanian culture oversaw the portfolio of the Ministry of Culture, nowadays it is politicians. Most of the time it's people who have nothing to do with research." From Alexandra's point of view, political appointees are less capable than cultural experts of appreciating the historical or scientific value of archaeological discoveries. Instead, political appointees are more easily persuaded by the promise of development, with its potential for adding tax revenues to lean municipal budgets, even when development compromises known archaeological sites.[32] The role of preservationist, Alexandra insinuated, is being played by pro-development appointees.

"The decision-making process is also becoming decentralized," Alexandra continued. As others have noted, the laws that created legal obligations to carry out archaeological analysis of development projects also empowered local officials to exercise their authority over the process.[33] "At first, a lot of archaeologists were happy because they thought it would be easier to make preservation claims by working with local authorities rather than through the Ministry," Alexandra explained. "But we quickly realized that this was not the case. We are completely beholden to local councils because that's where most museums get their funding." The promise of development resonated all the more deeply among local officials. Looking down at her coffee as she stirred, she continued, "There's a very real concern that if you don't do what the local council wants, then your funding is gone." With a dismissive shake of her head, Alexandra added, "And so now we have village mayors deciding what's important." From the view of practitioners, the process of adjudicating significance appeared impossibly stacked in favor of development. Even more troubling to Alexandra was that the process could be, in practice, easily sidestepped entirely. As others have noted, there are few instruments for enforcing patrimony protection laws.[34] "Local authorities can always just issue a construction permit without even consulting their county cultural department or regional commissions," Alexandra admitted. "Once they start building, there's nothing to be done." As our conversation drew to a close, she noted, "There are always ways of bending the laws to avoid a collision."

With unwavering dependability, processes intended to protect archaeological heritage from the fast-paced pressures of development instead certify their disposability in the name of it. The profitable gentrification of the underground demands as much. Bureaucratic processes, construction practices, and private funding provide market and municipal actors with reliable strategies to bulldoze through the underground's temporal obstacles by minimizing the likelihood of chance finds and by accelerating the glacial pace of stratigraphic excavation to hew as closely as possible to the timelines preferred by developers. While archaeologists wince over the perceived disregard for matters of historical and political importance, government and corporate entities reliably accelerate the slow pace of clearing the underground to meet the demands of development. And yet, as the next chapter recounts, the ability of market and municipal actors to maneuver, compel, and accelerate the expanded underground into being is not without limits. The unrelenting pressure to develop the city's verticality eventually causes it to buckle.

CHAPTER 7

═══

Foundations

T he confluence of foreign investment and rising consumption swirl-
ing around EU accession drove construction in central Bucharest to
new and dizzying heights. Since the turn of the millennium, foreign
investors have constructed ten office and residential buildings towering 70
meters or higher into the sky, three of which exceed 100 meters in height.[1]
More towers are planned for the years ahead.[2] Located predominantly in the
center-north region, soaring office and residential towers demarcate Bucharest's now buzzing central business district. SkyTower, the city's tallest
building at the time of writing, is emblematic of Bucharest's aspirations.
Stretching 137 meters upward, SkyTower markets itself as the capital's newest architectural landmark, one that trades not only in prime office space in
a central location, but also in the intangibles of image, identity, and status.[3]
To be sure, SkyTower's shimmering exterior and towering pinnacle dominate
the imagination when it comes to Bucharest's newfound verticality. Less remarked upon is the other part of SkyTower's jaw-dropping extension: its
cavernous depths. "We go a little more than thirty meters belowground,"
SkyTower's building manager Ruxandra explained from her office. "We have
five belowground floors. We made them parking because we can secure higher
office rents aboveground. Then there's another ten meters of foundation
and supportive steel piles beneath that," she spelled out, "and it wasn't
easy to build that deep, especially being so close to Floreasca Lake (Lacul
Floreasca)."

"Construction is undoubtedly deeper now than in the time of Ceaușescu,"
explained Luca, an engineer whose firm has carried out some of the deepest
excavations in Bucharest (see Chapters 5 and 6). "We've recently designed reinforced concrete foundations with structural elements that reach depths

Figure 23. Counterforce during excavation, 2019. Photo by author.

fifty-five meters belowground. The only real limit to how deep a building can go," Luca insisted, "is economic." But the engineer's initial assuredness glossed over the risks to developing downward that cannot be fully mitigated by simply upping the budget. Cracks and fissures have spidered across the city as office tower construction booms in what is already one of Europe's most densely built capitals. Investors are trying to get the most out of the city center's remaining plots of land by building not just upward, toward a higher skyline, but also downward into ever greater depths. "Most office and residential buildings have somewhere between two and five basement levels," Luca continued. These basement floors were usually dedicated to parking, but they also served a wide range of other uses: bomb shelters, grocery stores, and fitness centers, for example. "Underground construction at these depths poses significant technical difficulties; it's the part of construction that poses the greatest risk, especially in densely built urban areas, because the moment you excavate a pit in the soil, you risk collapsing any nearby buildings," Luca admitted. "Working at that depth takes a lot of careful calculation and oversight." When calculations are off or an excavator misses its mark, catastrophic accidents can occur that impact both the building under construction as well as neighboring buildings and roads, including the people who rely upon them.

The development of the underground, this is to say, brings the relationality of the city's architectural elements into clear relief. Weaving through Bucharest's winding streets, the city appears as a collection of discrete buildings separated by walls and gates as much as by deeds and plans. Shifting one's perspective belowground reveals a different kind of relationality to the city altogether. What once appeared separate—an office tower over here and a residential block over there—from belowground appear inextricably connected by the shifting grains of Bucharest's soft, sandy soil. In contrast to previous chapters, the obstacles to gentrifying and expanding the underground addressed in this chapter cannot be dispensed with through policing (Chapter 5) or bureaucracy (Chapter 6). Geology is not so easily overcome. Developers construct additional basement levels and even deeper foundations to anchor the penthouse views enjoyed by elites and to create additional space for the middle classes, but the new imperative to develop downward is not without its hitches. As the urbanist Paul Virilio might put it: the production of the underground is at the same time the production of its failures.[4] As became clear while working with developers, excavations at such depths can easily miscalculate the relationality of the underground, resulting in potentially catastrophic disasters. These accidents are not always neatly contained belowground, nor are they bound within a construction site. Instead, the perils of developing the underground reverberate upward and outward in ways that impact how others move through, inhabit, and enjoy the city. These negative consequences can occur when the underground is under construction (see part two of this chapter) as well as after construction is completed (part three). And once these belowground locations come to be inhabited, further risks emerge: underground spaces, rather than being safe refuges, are especially vulnerable to collapses, fires, and floods. As the key demographic occupying these spaces, the middle classes are therefore particularly vulnerable to such dangers (part four).

While the city's downward development allows for a new vertical ordering of the city, one that extends to the middle classes the dream of getting to live, work, and play in the city center, underground developments also risk the degradation of public infrastructure and the inconvenience of slowed commute times as well as the dangers to those underground posed by fires and earthquakes. While planners and developers turn toward the underground opportunistically, this chapter shows how suddenly these plans can become unstable.

Off Kilter

In 2001, the investor Antonis Kapraras and the Greek company Euroestate initiated construction on the Millenium Business Center in the center of historic Bucharest.[5] Billed at the time as Bucharest's tallest office building, the tower stretched seventy-two meters aboveground upon completion, and extended another twenty-eight meters belowground.[6] The project introduced an entirely new verticality to its surroundings, casting an unapologetic shadow over the low-rise buildings nearby, which dated back to the nineteenth century. As construction crews drilled and excavated their way downward to hollow out a pit for the Millenium's foundation, the changing forces within the soil caused the foundation of the neighboring Armenian Church (Biserica Armenească) to violently crack into several pieces. In an effort to stabilize soil pressure and thereby minimize further damage to the Church and other neighboring buildings, developers hurried to complete the Millenium's foundation.[7] Litigation soon followed, resulting in a US$790,000 settlement to stabilize and restore the church.[8] "Works at such a depth, in a sandy ground, in an area with high groundwater led to the strong cracking of the Armenian Church," a former Minister of Public Works and Territorial Planning commented in an interview with reporters. "These [types of constructions] seriously affect the urban fabric of the city and have particularly dangerous implications for the safety and stability of neighboring buildings, but also on green spaces, traffic, parking and water supply and sewerage systems." With towers proliferating in central Bucharest, the former minister concluded his remarks with frustration, adding bitterly: "It surprises me that since 2001 these concerns don't matter anymore."[9] While developers neatly contained their returns on their investments, the negative impacts of their construction upon public parks and utilities, roadway systems and commute times, and upon the city's parking supply were broadly endured, particularly by the middle classes who rely upon these infrastructures. Despite the concerns of the former minister about the Millenium and similarly scaled towers, the period following the controversy saw the approval of a raft of new office and residential projects for central Bucharest.

"The soil is a living creature," began Dinu, a professor at Bucharest's Technical University of Civil Engineering (UTCB), making an effort to be intelligible to the anthropologist with no civil engineering background. Dinu continued, "It's modifying pressure all the time because of excavations, seismic tremors, and also because of fluctuations in the water table. The soil has a lot of variability to it." Excavating tens of meters deep into sandy conditions, he explained, can trigger a resettling of the soil. If not appropriately managed, the soil's pressure upon existing structures can quickly change, and be accompanied by destabilizing effects. The problem, I was told time and again, is made all the more complicated by two factors. First, Bucharest's soft, sandy soil shifts easily. And second, Bucharest is densely built, which places deep excavations for new projects (such as the Millennium tower) cheek by jowl with existing buildings (such as the Armenian Church).[10] "It's not that conditions are impossible and that high-rise towers can't be built here," Dinu assured me. "But the conditions are atypical. It requires careful engineering."

The majority of Bucharest's historic buildings, Luca and others explained to me, have relatively shallow foundations of about two meters. This is because Bucharest's historically low buildings—of four to six stories in height—did not require elaborate foundations for their support. The recent rise in office and residential towers, by contrast, has introduced significant depth to the city and complex engineering beneath the surface. Engineers anchor foundations deeper underground to manage the increased weight of buildings, the shifting pressures of the wind above, the soil compressing around the building's base, and the seismic activity reverberating from way down below. "Heavy buildings on weak terrain cause compression, and compression of even a few centimeters can ruin your building or its neighbors," Luca told me. "Depth is one way to distribute the weight of a building to denser layers of earth." Luca noted that the verticality of New York is facilitated by the presence of bedrock between twenty and thirty meters beneath the surface. The search for bedrock in Malaysia to support the Petronas Towers' 451-meter twin pinnacles required piles that drive 114 meters downward.[11] "The taller the building, the softer the soil, the deeper its foundation tends to go," Luca summarized.

The process of digging deep foundations to stabilize one building, however, risks destabilizing the other buildings and roads nearby. The problem, Luca explained, is all the more pronounced in densely developed cities such as Bucharest. "The majority of office towers are utilizing the entirety of their land," he noted, "so excavation is occurring in close proximity to adjacent buildings and roads. Big problems like what were seen with the Armenian Church begin to arise once projects get beneath the second basement level." In 2018, Luca invited me to tour an office tower under construction in the city's center-north with Sergiu, a member of the Geotechnical Department of Luca's firm. "There is an entire science to predicting compression," Sergiu assured me as we walked the perimeter of the site. As we talked, Sergiu directed my attention equally toward the pit under construction before us and to the homes located just beyond the site's fencing. "Before the first scoop of dirt, we calculate the amount of compression given what we know about the project's plans and the site conditions, and then we plan accordingly." Sergiu gestured to a retaining wall designed to provide the counterforce they have predicted to be necessary to protect nearby buildings and roads. "But," he continued, "then we have to wait and see what actually happens in reality. If not properly accounted for, such large forces of compression pose serious consequences on the surface. If nearby buildings start to sink, then you've ruined them."[12] The same is true for roads as well as buildings, other civil engineers informed me.[13]

"There is fairly responsible oversight from the authorities so as not to impact neighboring buildings," Luca insisted at another meeting. "Authorized projects have an obligation to monitor the construction. We have to measure neighboring buildings and the terrain so that no significant impact occurs." To that end, before a single scoop of dirt is taken, Sergiu and his crew measure extensively. Each neighboring building is first solicited to be evaluated. "It's the most awkward part of the job," Sergiu chuckled, "because you're asking to enter someone's home to do a thorough check for existing cracks, and that includes searching their bedroom closets." From the interior of the bedroom to the exterior seam where grass meets the foundation, Sergiu's team identified, measured, and then catalogued each crack. They regularly returned to these neighboring buildings to document the emergence of new cracks or the extension of existing ones. Most of their attention, though, is directed downward. Sergiu described the installation of a grid system of "spider magnets" used to measure the earth's compression throughout construction. The crew drove a series of pipes around the project site downward into the soil, until the pipes were anchored into solid bedrock. Within each pipe were strings of magnets, shaped vaguely like spiders, spaced a half-meter apart. Compressions or heaving of the ground cause the magnets to move along the pipe's axis. "We establish baseline measurements before the project begins," Sergiu explained. "Then they start digging down to the first basement level and we take another set of measurements. As they dig downward to each level, we keep measuring until they've reached their desired depth. And then we keep measuring while the structure is built, while the finishes are loaded on, and while the building is in use." As magnets shift in depth, Sergiu is able to track changes in the soil brought about by compression as well as by seismic activity.

For Luca, the potentially fatal consequences of even a partial building collapse provided clear and compelling motivation to spend time and money thoroughly monitoring his own projects. Others in the business, however, insisted that Luca's diligence is exceptional. Geofri, a civil engineer who performs contract work for the municipality, underscored the reluctance of most developers to carry out the expensive work of evaluation. Although Romanian law requires systematic monitoring of buildings over ten stories every six months, Geofri spitballed that fewer than 1 percent of constructions actually complete regular evaluations.[14] "People are supposed to be closely tracking if the building sloped or if cracks suddenly appeared," Geofri explained, "but in Romania there's a sense that this kind of work is expensive and so investors avoid it when they can." During our conversation, Geofri pulled out a heavy binder containing his own collection of news clippings about partially collapsed buildings around town, some of which he helped to restabilize. He invited me to flip through the binder as he talked. "Sure, there are interventions when a building finally collapses or once a crater appears in the road," Geofri continued, "but surely this is a cost that could be avoided."

While market actors predictably tried to skirt regulations, the sheer number of new building constructions in Bucharest prevented thorough government oversight.[15] "Since 2000, Bucharest has developed chaotically," began Tyna, a geophysicist at the National Research-Development Institute for Earth Physics (Institutul Național de Cercetare-Dezvoltare pentru Fizica Pământului). Part of her portfolio entailed predicting how the city will perform during Bucharest's next major earthquake. "The state's construction inspector is completely overwhelmed by the rate of new building constructions," she continued, trying to be diplomatic. "They just don't have the capacity to inspect every building." Not surprisingly, developers stay on time and under budget by cutting corners while inspectors are not looking. While those invested in the project save time and money, the broader community incurs the costs when projects hedge too aggressively, causing the sands literally to shift beneath everyone's feet. The inability of government authorities to monitor buildings over ten stories troubled Tyna given the regular seismic activity reverberating from deep beneath Bucharest. "It's a problem because we can't confirm the integrity of particular buildings, much less of an entire neighborhood," she continued. Noting that collapsed buildings, whether from compression or seismic activity, invite injury and death, she added, "growing instability creates another dimension of risk that we just cannot calculate."

Washed Away

In 2007, construction began for the eighty-meter-tall Euro Tower next to Circus Lake (*Lacul Circului*) in the center-north of Bucharest. In addition to rising twenty stories above ground, Euro Tower extends five levels belowground and rests upon a thirty-six-meter deep foundation.[16] Nearby construction on a seventy-two-meter residential tower with a thirty-four-meter foundation of its own soon followed.[17] As construction crews excavated and poured the buildings' foundations, the water levels of Circus Lake unexpectedly dropped several meters in a matter of months.[18] A deputy director for the Lake Administration attributed the unprecedented loss in water level to the depth of constructions unfolding nearby, positing that concrete basements and steel foundations were blocking the subterranean springs that supply Circus Lake.[19] Nearby residents voiced concerns as the quality and character of their public park deteriorated. In 2009, to prevent Circus Lake from disappearing entirely, the municipality excavated a fifty-meter-deep well to sustain the water level.[20]

As Bucharest developed downward, the proliferation of larger and deeper building foundations and basement floors, new Metro stations, pedestrian passageways, and parking garages brought the city into increasing conflict with its shallow aquifers. "Bucharest's water table varies between two and nine meters belowground," explained Cristofor, a civil engineering contractor, during a meeting at his office. "That means when we work deep, we work under the waterline. And that doesn't bother me," he said assuredly. Leaning back in his chair, Cristofor spoke confidently about his technical capacity. "I'm an engineer. I can work a solution. But it is another complication to work around." To be sure, Bucharest's rising skyline evidences the success of Cristofor, Luca, Geofri, and others in building over and through the city's groundwaters. Their individual successes, however, belie the collective consequences of how each project obstructs natural groundwater flows. "Each of these developments creates barriers that disrupt the natural dynamics of the ground water," explained Radu Gogu, the Director of UTCB's Center for Engineering Research on Subterranean Water. "The problem," Radu continued, "is that developers aren't acknowledging that these barriers are impervious and that they are proliferating. And so nobody is taking the cumulative consequences of these blockages into account." As technically adept as each tower's execution may be, Radu argued that the relationality of underground construction has not received proper attention.[21] While individually each project was not particularly problematic, their cumulative affect poses implications for the city's surface that reverberate across property lines. Obstructing the regular flow of groundwater, Radu explained to me, shifts the hydraulic pressure beneath the city. Increases to the groundwater level can flood basements, place upward pressure on building foundations, and create craters in roads. Decreases to the groundwater level, meanwhile, impact surface water levels, destabilize buildings, and increase sewer leakages.[22] These negative consequences do not fall upon the investor or the developer of a particular project, but rather are collectively endured by the project's neighboring residents, businesses, and commuters.

The negative effects of building downward are further compounded by the city's aging web of buried water pipes. While basements and foundations create blockages that obstruct groundwater flows, aging pipes hemorrhage staggering volumes of water into the city's soil. During the time of this research, the company operating Bucharest's tap water system, Apa Nova, reported network losses of 42 percent.[23] "It's high but not that much worse than elsewhere in Europe," Radu contextualized for me. "There has been a big effort to repair the entire water supply system. In the 1990s, there were areas of the city where water network losses were at 80 percent or more." Alongside leaky tap-water pipes, the city's socialist-era heating provider, RADET, reported stunning water losses within its own aged network of pipes. RADET spilled, on average, 1,835 tons of water per hour during the winter months and 1,225 tons per hour during the summer.[24] Combined, this amount of water loss not only can produce flooding on the city surface, but also contribute to erosion beneath the surface, as water losses wash away the city's sandy soil. The erosion of subterranean soil creates instability in the city above, exacerbating the risk of road and building failures. "It's really hard to conceptualize," Claudiu, a manager of a RADET plant, explained as he tried to put his plant's water losses into perspective. We met at a cafe offsite so he could speak more openly. "We report our losses in tons per hour, but think about the accumulated effect over days, months, and years," he continued. With a growing tightness in his voice, Claudiu stated his concern: "It's why we've started seeing 'craters' [*groapă mare, crater*]."

In May of 2015, a crater measuring eight meters wide and four meters deep opened up next to the Izvor Bridge in central Bucharest. Journalists expressed relief that no car happened to be above the hole when it appeared, though it was large enough to swallow one and had emerged on a major thoroughfare. The Apa Nova water company attributed the crater to a leaky hundred-year-old sewer pipe running beneath the bridge.[25] The city imposed a wave of traffic restrictions so that roadcrews could make repairs, bringing already slow-moving traffic to a halt.[26] Within a few months, a similar incident had unfolded in the adjacent zone of Tineretului. There, damage to the water network again caused asphalt to swell and break, spraying water up and out onto the road.[27] Shortly after that incident, a boring machine used to dig the Metro tunnel for the new M5 line struck a water drainage well near the Eroii Revoluției Metro Station. The damaged pipe released a rush of water that quickly softened the surrounding soil, sending five hundred cubic meters of water, gravel, and sand down into the station below. The accident left a gaping hole, only a short distance beneath the main boulevard above. With fears that a massive crater would emerge in the middle of a major thoroughfare, officials evacuated residents from two nearby buildings and restricted traffic on the main boulevard for thirty days. Construction crews, for their part, poured nearly two hundred cubic meters of concrete in order to consolidate the hole and to support the road above it.[28] While these craters received press coverage because of their unusual size, Bucharest's roads have long had a bumpy reputation among residents for potholes and craters.[29]

"Our drivers like to complain about the quality of Bucharest's roads," Geofri, whose consulting firm made road repairs for the city, explained. "But these are not failures of the road itself, of the asphalt." Taking the craters that formed near Izvor Bridge and Tineretului as cases in point, he continued, "the principal cause for road repairs is water infiltration from cracked pipes. Leaky water sweeps away fine particles. Eventually, the soil erodes enough that the road falls down into the empty hole that formed beneath it." The instability of Bucharest's roads, Geofri insisted, had to do less with the quality of the road system's execution and maintenance than with the hemorrhaging water networks running beneath the city. "It's the sewers, it's Apa Nova, it's RADET," Geofri continued. "If water is washing away the soil beneath the road, it's to be expected that the asphalt above will cave in," he shrugged. Geofri was quick to insist that such incidences, though common in Bucharest, were not unique to it. "These happen wherever water, soil, and asphalt interact, even in your New York," Geofri directed defensively back at me.[30]

Taken together, the populating of the underground with bigger and deeper building foundations alongside an extensive and aging web of pipes have shown the underground to be a consequential site of relationality. Developing downward creates the conditions for negative consequences that reverberate back up upon the surface, extending beyond project sites and property lines in ways that can elongate commute times, compromise the integrity of apartments, and degrade the character of public spaces. While the profits that development reaps are well contained within project sites, the costs that these developments incur are endured collectively, most squarely by the middle classes who rely upon the impacted roads, buildings, and public parks.

Vulnerable

As market and municipal actors worked to move them into the city's gentrified and expanded underground, the middle classes have become particularly susceptible to the accidents that are endemic to subterranean spaces. Their vulnerability crystalized after the Club Colectiv fire on October 30, 2015, which killed 64 people and hospitalized 172 more (see Chapter 4). As would become clear in the months that followed the fire, the underground nightclub had operated with a reckless disregard for fire safety regulations: the Club was overcrowded, the basement lacked emergency escape routes, the ceiling and walls had been lined with flammable materials, and the owners had allowed stage pyrotechnics.[31] In hindsight, the accidental fire should not have come as a surprise. For fire inspectors though, the Colectiv tragedy raised broader concerns that rippled across the historic center, where investors were breathing new life into dilapidated buildings. "When a person renovates their building," Liviu, a fire safety inspector, explained to me a few months after the tragedy, "and they want to change the use of their basement, they need an authorization. We have laws about this—its's not like anything goes." By Liviu's account, the renewal of Old Town led to a spike in the number of authorization requests. Investors were interested not only in stabilizing and restoring the historic center's stately nineteenth-century manors but also in converting them into offices, restaurants, bars, and clubs. "A major component of the problem is that businessmen see a room of a certain size," Liviu continued, "and they think, 'I want to put a club there.' But that's the wrong question to be asking. They're not considering the original, intended usage of these spaces. They should be asking, 'Does the construction lend itself to being a club?'" Across central Bucharest, Liviu worried, businesses were utilizing basements for purposes they were never designed to accommodate. "People aren't waiting to get their authorizations in order," Liviu said, shaking his head. "Any space becomes dangerous when regulations aren't respected."

The skirting of regulations underground is particularly troubling, a group of firefighters explained to me, because a building's basement levels are more difficult to secure than their aboveground levels. Unlike aboveground spaces—from the ground floor to the penthouse—underground spaces have no exterior. "When we're aboveground," as one firefighter put it, "there's an outside to the building to work with. You can break windows to clear out the smoke, to move in and out." Basements, by contrast, are burrowed out of the earth. They offer no space in which to maneuver strategically. "It's not a phobia," another firefighter insisted. "The underground is just physically and psychologically harder to work, even compared to the very top level of towers. Buildings that take us up there have much more demanding fire safety systems, extinguishing systems. But when you step underground," he contrasted, "it's another mood entirely." His colleague agreed. "You can't clear

Figure 24. Union Metro passageway, 2018. Photo by author.

out the smoke. The visibility is so much worse. When you step underground you start to feel out of your element, like you're now a miner."

In addition to being more complicated to secure, underground structures have also had their very stability called into question by recent events. Experts had long assumed underground structures to be largely immune to the earthquakes and air raids that rattle and shake the city above in ways that can quickly turn buildings into rubble.[32] For that reason, bunkers in Bucharest are buried belowground and the subway system doubles as a civil protection shelter in the event of a catastrophe (see Chapter 9). At the time of the Bucharest Metro's initial design and construction, the ability of subway stations to weather a seismic event went unquestioned because they were assumed to be safely embedded underground. The resiliency of the Metro system in the face of an earthquake was not tested, nor were subway stations and tunnels closely monitored during or after construction, making it difficult to assess earthquake-related damage.[33] By the turn of the twenty-first century, however, the flaws in these assumptions were laid bare by the complete collapse of several Metro stations following powerful earthquakes in other parts of the world, most notably in Kobe City, Japan, and Chengdu, China.[34] These events have led Tyna and her colleagues at the INCDFP to question the ability of Bucharest's Metro stations and tunnels to withstand Bucharest's next great earthquake at the very moment municipal actors are working to move more and more middle-class commuters down into its network. "It's now clear that Metro station galleries can collapse," Tyna warned. One of the most important nodes within Bucharest's Metro system, Union Station, exists beneath significant building foundations as well as the Dambovita River; the station is also encased by the city's soft soil, which is prone to shifting in an earthquake.[35] "We don't know how these structures will ultimately behave," Tyna admitted, "but there is a real risk of trains derailing, fires starting, vestibules collapsing, and also of passageways and platforms flooding." At the time of our conversation in 2018, Tyna and her colleagues had discussed the risk of inundation at Union Station as an abstract possibility. Extreme weather events in subsequent years led to well-publicized instances of Metro station and tunnels becoming dangerously inundated in other parts of the world, such as in Zhengzhou, China, where floodwater rose to neck-level inside its subway tunnels.[36] The middle classes invited to commute into the city by way of the underground assume these risks every day, whether or not they are cognizant of them. The first responders charged with helping those caught up in such catastrophes acknowledge the very real possibility of accidents but respond, nevertheless, with bravado. "We're professionals and we train. But it's true—everything gets more complicated underground," one firefighter shrugged. His colleague nodded in agreement.

*   *   *

As the density of development overburdens the city, the expansion of underground urbanism has offered one way of clearing the surface. The production of subterranean parking garages, transit systems, and basement levels has readied the underground to absorb an increasing array of urban life's most functional aspects, allowing the surface to be staged for Parisian fantasies meant to be enjoyed by those who remain on top of the city. But, as the city develops ever more aggressively downward, the relationship between the city above and the city below has grown ever more complicated. While the underground can facilitate the aestheticization of the surface in the name of furthering economic growth, the production of depth also risks destabilizing the city's surface. Although located out of sight, beneath the view of the sidewalk, subterranean constructions make their presence known above, in cracked building foundations, craters, and vanishing lakes. And the structures themselves, long imagined to be safely ensconced underground, are now seen as vulnerable to catastrophic collapses, fires, and flooding. Amidst the optimism of development, as more and more aspects of urban life go underground, one observation has become abundantly clear: the city's vertical ordering can quickly turn unstable.

CHAPTER 8

Digital Public Library

"We wanted to pull Romania into the digital era," Raluca explained in the summer of 2015, her colleague Anca nodded along in agreement. We sat in a conference room in the Bucharest headquarters of the global advertising firm McCann. Raluca and Anca were part of the creative team behind an evolving advertisement campaign entitled the Digital Public Library (Biblioteca Digitală, or DPL). First introduced in the fall of 2012, the campaign paired the prominent telecom provider Vodafone with the Romanian publishing house Humanitas to produce an interactive media installation inside of the Victoriei Metro Station.[1] Deliberately framed as a public library (*bibliotecă*) rather than a bookstore (*librărie*), the DPL manifested visually as floor-to-ceiling wallpaper, which gave the impression of standing before a meticulously (perhaps impossibly) well-maintained continuum of bookshelves. Each shelf appeared to be stacked tightly with hardback books of the exact same size, their spines facing outward. Printed onto the spines of each of the library's five hundred shelved "books" were titles from prominent Romanian authors alongside canonical works by the likes of C. S. Lewis, Franz Kafka, and Milan Kundera. The visual effect was undeniably striking. McCann designed the DPL to be brand agnostic, allowing any Metro commuter with any smart device, regardless of their cellular service provider, to scan the QR code associated with their desired title and download a free sample chapter. Passengers-turned-patrons then had the option to purchase and download the remainder of the book, thus becoming consumers. "It was wallpaper," Raluca explained, "but you know, it was like the public library as you know it—just brought into the digital era." Of course, practically speaking, the installation had much more in common with a private bookstore than a public library, but Raluca's framing spoke to the

Figure 25. The Digital Public Library, 2014. Photo reproduced with permission from Paginademedia.ro.

project's ambitions as a large-scale cultural endeavor, and as an exercise in molding and shaping a broad public community underground. The DPL received both industry and public acclaim, winning several international advertising awards and inspiring copycat installations in London, New York, and Beijing.[2] Adopting the tagline "a library for every student," the DPL expanded upward and became an integrated part of the Romanian education system, its reach extending into some three hundred public high schools.[3] "The Metro just fit so nicely into our strategy," Anca commented as though it was self-evident.

This chapter takes the Digital Public Library—its location inside the Metro as much as its developers' civically minded commitments—as an opportunity to shift this book's focus toward a consideration of how public life extends underground. The endeavor is long overdue. Free public libraries originally developed alongside the modern, democratic ideal of urban space, helping to prepare liberal citizens to participate in the public life of the street.[4] And yet, as Lewis Mumford noted while studying the formation of the great metropolises of the early twentieth century, "beneath the visible city an invisible city grows apace . . . a city of ramifying subways and ominous tunnels in which the entire population spends no inconsiderable part of the day."[5] As market and municipal actors have developed the city downward, more and more people find themselves spending larger amounts of time beneath the street than upon it. Despite the installation's eye-catching execution, and advertisers' insistences that it would bring "the public library as you know it" underground, this chapter argues that the DPL, and the gentrified underground in which it took root, offers only a flattened experience of the modern ideal of public life, one that is more individualized, less social, and more commercially mediated.

Disconnected

"We wanted people to do something positive with the time stuck waiting on the Metro," Raluca began cheerily as we discussed the inspiration for the DPL, "to fill up the twenty or thirty minutes that are otherwise wasted." The Metro may very well be a site of heightened locomotion, where bodies are shuttled between home, work, and leisure beneath bottlenecked roadways, but, as Raluca insisted, riders tend to experience the Metro just as much in terms of feeling stuck waiting. Passengers are stuck on platforms waiting for the train only to be stuck on a train waiting for their station. The sequence resets and repeats for those needing to change lines. The DPL's underlying assumption, Raluca explained, was that these feelings of being stuck waiting and of time wasted were characteristic of the Metro system itself.[6] Positioned beneath the city's surface, the Metro is physically disconnected from the public life of the street. In the modern ideal, public life is characterized first by open space to be used and enjoyed by everyone, and second by forms of consumerism that are accessible to all, even if only as a visual spectacle to be taken in and contemplated in passing.[7] This ideal of public life is characterized by the effervescence of the open street as both a space of encounter and of belonging. The public life taking shape within the confines of the Metro, by contrast, had so far failed to embody these same possibilities of social connection and bustling vitality. Coffee kiosks proliferated underground, for example, but the liberal ideal of coffee houses as centers of cultural and political criticism did not.[8] As the case of 5 to Go illustrates (see Chapter 2), the rhythm of the underground constrains baristas' rapport with customers rather than encouraging it, and kiosks offer no seating for customers to discuss (much less debate) the news of the day. Prior to the DPL campaign, furthermore, the free public libraries that educated citizens in how to belong to both the street and the nation were entirely absent from the underground.[9] Without the foundational institutions of public life, the Metro's social and material characteristics failed to capture, much less stimulate, the attention of its passengers as they waited.

"There's just not much to look at," Anca bluntly insisted. Bored out of subterranean soil and rock, Metro stations are strikingly discrete spaces, offering no signs of life beyond their boundaries.[10] In sharp contrast to the bus stands that dot the city above, for example, Metro stations provide no view onto a city square or boulevards beyond, whose million tiny dramas, architectural splendors, and commercial spectacles have long been a source of idle stimulation.[11] These elements of the public life of the street are, Walter Benjamin noted, not without a sense of intoxication for those who choose to look upon them, contemplate them, and decipher them.[12] Yet Metro stations, as subterranean enclosures, extinguish from view the city's many provocations and pleasures, diminishing the vibrancy of urban places, as others have put it, to "dots and colors on a map, or block letters on the station walls."[13]

Just as important, from the campaign's perspective, Metro stations were also disconnected from the digital networks that were then expanding across the city and the country. "Before the campaign," Raluca started, shaking her head disparagingly, "there was no signal whatsoever underground—texting, much less downloading data, while in the Metro was unthinkable." To be sure, the shortcomings of cellular networks left Metro passengers without service on their commutes. However, the obvious shortcomings also aggravated lingering suspicions that Romania's quality of life lagged behind Western Europe.[14] The lack of cell signal, Raluca and Anca as well as Metro riders themselves insisted to me, indicated that they were not participating in a contemporaneous modernity with the likes of Paris and Brussels, a modernity marked by the seamless integration of mobile phones, digital maps, and texting into their everyday life.[15] Cut off from the street and disconnected from digital networks, the Metro system's immediate interiority set the stage for the passenger experience. The staged materiality of the Bucharest Metro did little to engage its captive audiences. "We don't have cool Metro stations," Anca continued dismissively. "The Metro hasn't improved since the day it was built," Raluca chipped in. For the hundreds of thousands of passengers who found themselves held in limbo underground as they waited, the Metro's stations were in no condition to occupy the attention of their passengers.

Alongside the stations' muted design, intensive regulation of Metro passengers prevented the creative appropriation of space—of station vestibules,

Figure 26. Rush hour, 2017. Photo by author.

platforms, and passageways—that gives street life its improvisational character.[16] Unlike the sidewalk, the Metro system's traversed spaces are not public and freely accessible. After becoming a refuge for unhoused "street children" in the 1990s, Metro administrators initiated aggressive policing against loitering and panhandling within its system.[17] Ticketing policies were enforced at each station via turnstiles overseen by a combination of ticketing agents, police officers, and private security guards. Police officers also patroled the busiest stations, such as Union, Gara de Nord, and Victoriei; private security guards were stationed on most trains to move along informal vendors, solicitors, or hustlers. While such policies have allowed middle-class passengers to travel unaccosted, they have also dulled the pleasure of encountering people we experience as different.[18] Such policies foreclose the emergence of "public characters": the chatty retirees, street musicians, pastors, or vendors who idle in public, and, in so doing, give public spaces their sense of vibrancy and effervescence.[19]

While milquetoast stations and intensive policing muted public life, the high volume of passengers—particularly at rush hour—complicated mental escapes into work or entertainment. Even as the municipality called for a greater reliance on the Metro, packed trains squeezed passengers cheek by jowl during their commutes. Riders bumped into one another as trains rattled down the track. The jostling only intensified when doors opened and the acts of boarding and exiting had to be quickly negotiated. A mild slowdown in service could overwhelm Metro stations to newsworthy proportions while brief stoppages at times required police officers to manage the foot traffic on overwhelmed platforms.[20] Under such a density of bodies, cracking open a book, much less adjusting a spreadsheet on a laptop, was practically impossible.[21]

In the enclosed and highly regulated space of the Metro that left the middle classes feeling stuck while they waited, perhaps no aspect of public life flourished more than advertising. "In the Metro, advertising is everywhere," Mariana, an advertising executive at another firm, explained. "It's the ideal medium," she assured me. A chuckle growing in her voice, she continued: "People stand on the platform and look at ads because there is nothing else for them to do down there. They can't change the channel; they can't look down the road." Having inadvertently assembled a captive audience, Metrorex has cashed in on its commercial potential by making every square inch of its system available for rent.[22] "If you have the money," Mariana insisted, "you can brand anything down there." Walking through any major Metro station at any given moment evidenced as much. In Union Station, for example, advertising lined designated billboard spaces along station walls; banners, decals, and graphics pushing bathing suits, food delivery services, and of course cellular credit covered the columns, floors, and stairs of train platforms. Advertising's reach even extended onto the windows, floors, and straps of the trains that connect stations. The Metro's willingness to allow any of its surfaces to be covered with branding has made possible wondrous, three-dimensional, and immersive advertisement installations that transform passengers' experience of the Metro's space and its public life.[23] The Vodafone Digital Public Library is a prominent case in point.

Digitizing the Underground

McCann's Digital Public Library concept took shape at the intersection of corporate investment and post-socialist anxiety, aiming to make Bucharest materially and culturally more comparable to Western Europe. Vodafone, to explain, had tasked McCann with designing a campaign to do two things, the first of which was to announce to Romanians their incorporation into the so-called "digital era." This ascension, from Vodafone's perspective, was made possible by its €500-million investment to upgrade and expand its data network infrastructure in Romania. The investment, Vodafone argued, fundamentally transformed Romania's lagging cellular network into one that offered customers the fastest and most comprehensive coverage in all of Europe.[24] "At first in the 1990s," Raluca recalled with disapproval, "there was no signal whatsoever." Calls regularly dropped, and building elevators, stairwells, and the entire Metro system were dead zones, she insisted. Vodafone's campaign was meant to show Romanian consumers that they were now connected to the Internet anywhere, from beneath the capital to across the Romanian countryside.

The campaign's second aim was as economically driven as its articulation was civically minded. It needed to make Vodafone's investment profitable by habituating Romanian customers into consuming bandwidth at rates more comparable to elsewhere in the EU. Mobile data consumption in Romania was, at the time, only a third of the European average.[25] Vodafone had an obvious interest in better incorporating mobile phones into everyday activities. After significant brainstorming, McCann built its campaign around the act of reading. "Reading is this classic activity people already associate with riding on the train," began Raluca. "We liked the idea of taking books into a digital environment as a way of making reading available to people when they need it, when they're otherwise sitting bored on the train." McCann explicitly framed the campaign's commitment to reading in terms of self-betterment. "We didn't want to just provide entertainment—it's not about that," Raluca insisted. "We wanted to give people more—the time to learn something new." Anca nodded in assent. Romania's struggling publishing industry, for its part, welcomed the added exposure that McCann's e-library concept would bring. At that time, a string of studies had denounced Romanians' low reading rates compared to those elsewhere in the EU.[26] Public handwringing had ensued, raising self-conscious questions about the depth of culture and education in Romania.[27] With its commitments to connecting Bucharestians to digital, financial, and even cultural networks, the DPL campaign was as much an exercise in building and shaping a public as it was in cultivating Vodafone's brand image.

In the face of such social and cultural anxieties, the DPL campaign sought to foreground Romania's technological advancement and to build a renewed excitement for reading from the underground upward. "We wanted to demonstrate that the network has evolved," Raluca assured me between sips of her cappuccino. "Reception is now just as good anywhere here as it is else-where in the world." Besides facilitating communication, the comprehen-sive rollout of such digital technologies has become intertwined with global ambitions, linking individuals equipped with mobile phones to possibilities that exceed both their immediate and their national bound-aries.[28] The underground, as a space completely manufactured by modern engineering, emerged as the ideal site for heralding Romania's entry into a new techno-future.[29]

McCann's demonstration of capacity was physically installed at the in-terchanges of the M1 and M2 lines at Victoriei Station. The campaign's loca-tion enabled office workers commuting from new residential blocs along the periphery to download something to read on the second leg of their trip up-town, toward the office parks in Pipera, or on their final connection at the end of the day. The DPL's canvas stretched from floor to ceiling in order to literally paper over the aging station walls, which had not been built to im-press. In sharp contrast to the Moscow Metro's museum-like elegance, the Guastivino arches of early New York City subway stations, or even the do-mestic comfort of the initial London tube stations, Bucharest's socialist-era planners did not invest in such aesthetic embellishments.[30] Instead, in the name of establishing a more egalitarian society, Bucharest's designers had fo-cused upon the Metro's functional capacity to move tens of thousands of riders a day across the city and toward a new image of the future.[31] In its new-est and brightest state, commentators expressed pride at the Bucharest Met-ro's pragmatic modesty, and its emphasis on convenience over grandstanding.[32] And with Victoriei Station's appearance suffering from decades of wear and underinvestment, Anca and Raluca framed the act of papering over the en-tirety of a wall as a civic act of beautification.[33] Left unstated were two other aspects of the DPL's impact: its emphasis on the underground's aesthetic form over its function, and on providing individualized and commercially medi-ated transit experiences over servicing a collectivity of riders.[34]

The installation of the Digital Public Library fundamentally transformed the experience of the space. The Metro's aging, modernist minimalism gave way to the impression of standing before a wondrous continuum of wooden bookshelves. Each tightly stacked row of identically sized hardbacks had a dominant color palette—reds, blues, yellows—on which each "spine" offered variations. Every shelf ended with a decorative flourish meant to disrupt the potential monotony of such a perfectly clean composition as well as to pro-vide an added sense of depth: a vase at the start of one, neutral-colored books stacked three spines high at the end of another, the dust jacket of a promoted title facing outward in the middle of another. At the top of the fifth row of books, beyond the reach of any adult, a simulated catwalk appeared printed over rows six and seven of the library, giving the impression that the installation had a second floor.[35] As intended, the effect piqued people's curiosity. Commuters, Raluca recounted with a smile, inquired if there were different titles that were only available up there, just out of their reach.

In its totality, the Digital Public Library's composition was undeniably pleasant to look at, as much for the warm bookcases that its canvas simulated as for the muted station walls that it, and other similarly ambitious campaigns, covered up. The spectacle of immersive advertising provided commuters underground with their primary reprieve from the brute materiality of the Metro's labyrinth of dingy white and mildew yellow tile-encased tunnels and galleries. From an advertising perspective, the underground's uninspired characteristics were also what made the Metro an ideal backdrop for forming a particularly productive kind of public, one understood to be marked as much by a shared physical space as by a presumed sense of boredom.[36] "At the metro, people just stand around waiting on the platform, which doesn't give them much to look at: no clouds, no buildings, no traffic," advertising executive Mariana explained as though telling it to me straight. "And so that's why we put billboards up in the Metro. I don't even think people are consciously reading the billboards," she shrugged, "but the messaging enters their thoughts as they stand around and wait." In a subterranean landscape deprived of what Walter Benjamin described as the phantasmagoria of nineteenth-century Paris, the intoxicants of advertising—however unappreciated by those passing through—constitute a kind of shared text.[37] Advertisements like the Digital Public Library have helped to transform a constantly changing aggregation of Metro riders into a durable public capable of receiving marketing suggestions.[38]

Forming the Reading Public

The DPL's stunning scale and execution aimed to aesthetically repurpose the Victoriei Station Metro interchange into a digital articulation of "the public library as you know it" (in Raluca's words), one that formally espoused a commitment to "free education" mediated by mobile technology.[39] Yet as scholars of libraries have noted, the common understanding of the term "public library" has varied significantly over time and space.[40] The difference between the public at large and the reading public that a library serves can be gleaned from the kinds of books and the forms of reading that it facilitates. Five elements of the DPL's form and catalogue composition provide a glimpse into the kind of public the DPL called into being underground.

Most obviously, the DPL's holdings were exclusively books: they did not include other kinds of media that are typically circulated at public libraries, from magazines and comics to music, videos, and general internet browsing. The decision to offer only books is worth considering given that the technical demands of music, videos, and web browsing would have provided a more intuitive demonstration of a digital network's capacity than downloading e-books, which are singular and relatively small files. While not denying this point, Raluca and Anca framed the DPL's narrow focus upon books as civically minded. "You can listen to music anywhere," Anca deflected, "but that isn't going to build you as much as reading a book." The impetus to build the public up via reading coincided with a European Commission effort to promote democracy and active citizenship in Romania and beyond by supporting digital services across Europe's public libraries.[41]

The DPL's curated catalogue of some 500 book titles, second, focused squarely upon canonical texts rather than a diverse selection of popular reading.[42] The DPL invited passengers to download and skim on their crowded morning commutes the signature works of the likes of Herman Melville, Immanuel Kant, and Fyodor Dostoevsky. Some of these volumes were in English rather than in Romanian. Placed alongside these canonical works—digitally represented on "spines" of equal size, thickness, and font—were titles by prominent Romanian authors, such as the national poet Mihai Eminescu, the novelist Mateiu Caragiale, and the historian and literary critic Nicolae Iorga. With its commitment to the canon, the DPL promoted what might best be described as "solid literature" intended to cultivate public virtue.[43] Every bookshelf, Pierre Bourdieu insisted, is indicative of a "market."[44] Rather than curating a diverse catalogue that sought to reach as broad of a readership as possible, the DPL focused its attention upon the educated classes seeking reinforcement of their self-assurance through their ability to aesthetically appreciate the cannon.[45] The DPL's tie-ins with Nobel nominee for literature Mircea Cartarescu and the Romanian public high school system reaffirmed as much.[46]

The digital element of the DPL, third, made participation in this public library contingent upon ownership of certain digital technologies. "For our purposes," Anca told me, "the device was arbitrary. We wanted to show that the network works with everyone's devices: Kindles, smartphones, tablets and stuff like that." Inclusive in its brand, the library's dependence upon cellular-enabled devices nevertheless left out a considerable swath of Metro passengers who could not afford such personal technologies, were unfamiliar with them, or were physically unable to use such devices: passengers from the lower middle classes, the elderly, and riders with certain disabilities, for example.

While framed via a commitment to rigorous reading, fourth, the DPL was designed more specifically to promote books as digital objects that could be owned. Many of the canonical works listed in the DPL catalogue, for example, had open access editions where the DPL could have directed its users. The DPL could even have organized its catalogue entirely around open access content, allowing users to freely read through its holdings in their entirety (not to mention that Bucharest's technologically savvy students were quite capable of downloading pirated books anyway). Each of these adjustments would have affirmed the DPL's nominal commitment to reading as such. Instead, the DPL used sample chapters to encourage e-book purchases from a designated publisher. The paywall, from Anca's perspective, was part of the DPL's effectiveness. "With our campaign," she explained, "we wanted people to feel like they own the technology. They didn't have to worry about the quality of the network anymore. We wanted to let people know they could now focus on consuming content." Designed as an introduction to e-commerce, the DPL in practice functioned more like a bookstore than a public library.

The DPL, finally, promoted only a single kind of reading—solitary reading—to the exclusion of other forms. While solitary reading has become predominant in the modern era and is often assumed to be reading's only form, ethnographers of reading are quick to remind us that reading is a culturally and historically determined practice.[47] The Greeks and Romans nearly always read aloud rather than to themselves, and reading in ancient Jewish culture was also an oral, social, and collective act.[48] The groundwork for a Western private reading culture did not take shape until St. Augustine.[49] This sort of solitary reading hinges upon an architecture of privacy that moved from the monk's cell to the scholar's study and on to the modern experience of being "alone together" in places like Metro stations.[50] Not supported by the DPL were collective forms of reading found in traditional libraries, such as author readings, children's story times, or book clubs, for example, that promote collective social ties by way of a shared text. Unlike traditional libraries, the DPL did not have the staff, community meeting space, or even the social programming needed to provide a context in which to foster col-lective forms of reading.[51] After downloading a selection, the DPL left its users to read to themselves while they finished their commutes.

The DPL did not attempt to cater to the public at large by providing free access to the widest possible variety of cultural materials to people of all ages and backgrounds, as the public libraries that served as its nominal inspira-tion strive to do.[52] Instead, the DPL focused its resources on the reasonably tech-savvy, well-educated students and professionals with the means to pur-chase their leisure reading (along with a device on which to enjoy it). Once downloaded, DPL content opened up the possibility for a solitary reading ex-perience that could provide those on an over-crowded train with a glimpse of the feeling of being alone, even as they were squished together. The public called into being by the DPL connected with the kind of upwardly mobile office worker that the Metro, as much as Vodafone, wished to attract, over and above the broader public that it served daily.

Connected

By way of the Vodafone network, the DPL demonstrated to those equipped with the right devices that they were no longer cut off from the city above while underground. To the contrary, the technologically capable and up-wardly mobile were now connected to possibilities that exceeded both their immediate and national boundaries, even as they whirred beneath the city's overcrowded streets. "This is what was so obvious and what we wanted to build upon," Raluca explained, her enthusiasm building. "The Metro is a place where you really need to be connected. People use the Metro to get some-where—to work or to school—and they need time to prepare, or to let people know they're running late. With the campaign, we signaled that this was now possible." Like the canonical texts that make up the DPL catalogue, the con-nectivity offered by the digital underground was intended for the seriously minded. The campaign heralded the activation of the formerly passive, by em-powering ambitious passengers to become smarter students and workers as they commuted by simply consuming more bandwidth. The expanded cel-lular network also demonstrated how readily Romanians could buy into the feeling of being incorporated—culturally as much as technologically—into a globally integrated modernity imagined as already being experienced in Paris and Brussels.[53]

Figure 27. The Digital Public Library, 2016. Photo by author.

As the campaign evolved over subsequent years, smaller "book shelf" in-stallations located inside traditional billboard spaces on waiting platforms declared the Bucharest Metro to be the largest free internet zone in Europe. The public inside Bucharest's digital underground was not merely fully con-nected to the world but now leading it. "This was a big statement and a big challenge to make that happen," Anca chimed in. "Now the Metro is a place where people use their smartphones all the time," she insisted. "If you're not on your phone you stand out. You'll look strange because you'll be the only one looking around." Rather than a space of encounter across difference, the public underground is a place where people stand out if they are not turned toward their own device. Riding the Metro confirms as much. In the name of connectivity, the overwhelming majority of Metro passengers are now pre-occupied with their devices: eyes fixed on flickering screens, ears plugged with headphones, and thumbs drumming on glass. These devices open up technologically mediated practices of avoiding others despite the Metro's imposition of spatial proximity.[54]

The kind of public life the DPL called into being underground was ulti-mately as flat as the wallpaper upon which it was printed. Despite its civically minded framing and formal commitment to "free education," the DPL stood contrary to the openness and inclusivity that have been the hallmark of modern public life on city squares and boulevards. Simply put, the DPL lacked the depth and resources to foster connections between people by way of texts, as traditional public libraries in Romania and beyond strive to do. Instead, it appropriated the minimal veneer of a library only to invite its public to tune out those who surround them and turn inward toward a private, personal reading experience made possible one purchase at a time. The DPL's move toward a public that is ever more privately curated, ever more antisocial, and ever more commercially mediated has much in common with the personal automobiles and enclaves that threaten the character of public space along the city surface by reinforcing boundaries and discouraging encounters with different people.[55]

Rather than fostering an inclusive public for those moving underground, the Digital Public Library instead succeeded first and foremost as a branding exercise for the Metro and the office parks that it services. It conjured for the upwardly mobile an appealing image of the Metro, one where young professionals navigate underground stations and passages with the distinction and cultural capital that the DPL instilled in the very public it created. The DPL reassured mid-tier professionals that they could ditch the privacy of their cars for a technologically mediated, personal experience on the Metro that would affirm their sense of class and status rather than draw it into question. As the middle classes commute underground, they encounter a public life whose quality and character take their cues not so much from the street and the square, or the traditional public libraries and coffee houses that line them, but rather from a mall that primes passengers to be consumers right up until they exit the Metro for the back office, where they are then squeezed for their labor.

CHAPTER 9

Bomb Shelters

At the very bottom of the city, tucked away in the back corners of the deepest basement levels of hospitals and office towers, kindergartens and universities, hotels and apartment blocks, government buildings and private banks, exist some 1,088 civil protection shelters.[1] Each and every one of these shelters is subject to standardized requirements intended to ensure the short-term survival of their inhabitants through the catastrophic collapse of the city above them. Their walls, made of reinforced concrete, must be a minimum of forty centimeters thick.[2] They are equipped with exit tunnels, also made of reinforced concrete, that stretch beyond the surface of the building to a distance no less than a third of the building's height to allow survivors a clear egress in the event of a building collapse.[3] Their ventilation systems are equipped with filters to remove toxic pathogens and radioactive substances resulting from an aerial bombardment, terrorist act, or industrial accident.[4] "We use thick concrete walls, special doors, and filtration systems to isolate the shelter from the rest of the building," Victor explained in 2018 as we toured one of the civil protection shelters he recently had completed for a new office tower. "They are equipped to handle a gravity load of five tons per square meter—about ten times the usual load," he said as he ducked his head to pass through an opened blast door. For all of the effort put into fortification, the interiors of these bunkers are strikingly bare—almost absurdly so. The law requires each shelter to have ten liters of water per person as well as a waterless toilet for every fifty persons.[5] They are otherwise entirely empty. The emptiness of Bucharest's civil protection shelters cannot be taken for granted as they contrast sharply with those of other countries. Switzerland's sprawling system of civil protection shelters, for example, is equipped to sustain inhabitants for weeks (if not months) after a disaster;

Figure 28. Civil protection shelter, 2018. Photo by author.

while, unsurprisingly, the privately financed bunkers of the U.S. suburbs are typically so well-provisioned with food, furniture, and entertainment as to be characterized as a "commodified dream space rather than a disaster zone."[6] Bucharest's civil protection shelters, for their part, have neither the food nor the medical equipment that would be necessary to hide away from a dystopian future befalling the city above for any significant period of time. While meant to materialize the dream of security in a precarious world, Bucharest's shelters are, by most accounts, of questionable utility in the post–Cold War era.

While their practicality may be debatable, civil protection shelters nevertheless remain an active part of public life in Bucharest.[7] The law requires that shelters be incorporated into most new building constructions, and that existing shelters must be maintained. Such shelters can be found in every sector of the city. The Department of Emergency Services (Departamentul pentru Situații de Urgență, or DSU) not only maintains a public list of these shelters, but also has created an app that maps their locations for anyone with a smartphone to find in a true emergency (assuming that cellular towers have not given way).[8] Given the tremendous sums of money invested in the production and maintenance of fallout shelters, as others have argued, their quality and coverage should be read as a window into the meaning of citizenship and belonging in the present moment.[9] The effort put into shelter standardization and their broad integration throughout Bucharest resonates with common liberal (as much as socialist) ideals of shared citizenship. However, Bucharest's ongoing commitment to civil protection operates within two basic limitations. The first is that there is only enough space for one million of the city's more than two million residents.[10] The brute, material fact is that, in a time of crisis, not everyone can be saved by civil protection shelters.[11] Not surprisingly, the very wealthy are the most likely to live and work in buildings equipped with shelters, while the very poor are least likely. The second limitation, closely related to the first, is that while these shelters are publicly mandated, they are privately produced and maintained at tremendous cost. "All of the reinforced materials that go into these shelters are expensive," Victor insisted. "The cost per square meter of underground construction is three times higher than the structure aboveground." Iosif, a retired member of the fire brigade who inspected civil protection shelters for decades, placed the cost differential even higher—somewhere between four and five times more expensive than the aboveground structure. "The architects and engineers forget about maintenance," Iosif insisted in defense of his higher estimate. "The air filters alone represent very big money. They require maintenance, they eventually need to be replaced, and they all need to be imported." To be sure, high-end office and residential towers keep their shelters impeccably to code.[12] "I never hear foreign investors complain about the cost of these shelters," Victor told me. "Investors only complain when they think they've found a corner that they can cut. But civil protection shelters really are mandatory expenses." The shelters beneath the city's aging, socialist-era housing blocks, by contrast, are maintained to varying degrees of inhabitability. The fact is that their middle-class residents cannot always afford to prioritize the proper upkeep of their shelters.

Bucharest's ongoing commitment to civil protection, this chapter argues, throws the limits of the middle class's incorporation into the public life of the city into biopolitical relief. In the absence of a shelter system that assures an "equality of survival" for all, Bucharest's middle classes occupy a precarious middle ground.[13] For the wealthy, the costs of installing and maintaining government-mandated civil protection shelters are relatively trivial, while for the poor, this is simply an unaffordable luxury, leaving them effectively abandoned by the state in a moment of crisis. The middle classes, however, are extended an unequal type of inclusion. Rather than being merely abandoned, the middle classes are extended a limited kind of incorporation: mandatory shelters pay lip service to their protection, but only to the extent that they are willing to divert money from already overburdened household budgets to maintain a shelter of questionable utility.[14] Against the dreams and aspirations of the ascendant middle classes, the city's broad variations in bunker availability and quality ultimately reveal the erosion of civic obligation and the degrading of social belonging in Bucharest. Just as not everyone can be middle class, not everyone is equally worth saving.

History of Civil Protection

The mass production of civil protection shelters in Bucharest originated in the early twentieth century, as Europe's great wars brought about the militarization of everyday life. The development of chemical weapons during World War I, and the introduction of air forces to deliver those weapons across once-unthinkable distances, dissolved any distinction that may have existed between home fronts and battlefields.[15] As the project of civil protection emerged in response to the evolving threats of war, those in charge have had to

Figure 29. Civil protection shelter inspection, 2017. Photo by author.

ask and answer anew two basic questions: who is responsible for ensuring protection and who (or what) receives full protection? On August 15, 1916, the Prefect of the Capital Police provided the initial answer to these questions in his ordinance "Measures Against Air Attacks," which he issued in the major newspaper *Adevărul*.[16] The ordinance established a set of civic obligations intended to reduce the number of victims from aerial bombings.[17] In this initial articulation, civil protection hinged upon the responsibility of citizens to keep each other and the city safe (the government did not imagine the countryside and its inhabitants as potential targets). Should an alarm sound, for example, the ordinance instructed residents to camouflage their own lights in an effort to render the city less visible from above. Only after darkening their homes were people instructed to take refuge in the already existing basement (*subsol, pivnițe*) of their building. Retreating to the basements and cellars of residential blocks, the ordinance reassured, would provide protection from bombs and projectiles as well as from building collapses. The ordinance also obliged residents to maintain a deposit of water for putting out the fires that would no doubt burn long after any raid concluded.[18]

During the interwar period, civil protection took on a more sophisticated shape. In 1933, High Royal Decree no. 468 and the French-inspired

Passive Defense Regulation shifted the responsibility for civil protection from the citizenry to the Ministries of the Interior and of National Defense.[19] Instead of directing citizens to douse their own lights and to seek shelter in their already existing basements, the state began constructing public air raid shelters. These shelters represented an entirely new kind of national infrastructure located completely underground. The state produced different kinds of shelters, modulating the level of fortification (and financial investment) in relation to the importance of the perceived targets.[20] Category I shelters, for example, were to be constructed in major city centers to protect industry and government. They were to be made out of masonry, reinforced concrete, and covered trenches. Category II shelters were intended for the public and were to be similarly made of masonry, reinforced concrete, and covered trenches. Category III shelters were to be installed in populated centers of major cities whose economic, military, or strategic importance placed them at risk of being targeted. Authorities equipped shelters across all three categories with either air filters or gas masks to protect against toxic weapons. Removed from centers of government and industry, the state cast residents of minor cities and rural populations as unlikely targets of an aerial campaign, and therefore not in need of state protection. Instead of installing civil protection shelters in the provinces, the state instructed rural citizens to protect themselves by digging their own "family shelters" in the form of a trench, basement, or cellar.[21]

The onset of World War II only heightened the social and political imperative for constructing more and sturdier shelters. Between April 4 and August 18, 1944, hundreds of U.S. bombers pounded Bucharest's railyards to disastrous effect.[22] "Our air attack on the marshalling yards at Bucharest was a bloody affair," recalled a U.S. Air Force commanding officer after one of his runs. There had been a public drill just an hour prior to his bombardment, which led many citizens to disregard the sound of warning sirens as nothing more than a continuation. "We killed about twelve thousand people," the officer estimated.[23] While the Anglo-American air raids focused on Bucharest's transportation infrastructure, they nevertheless inflicted significant damage to non-military targets. The high casualties of these raids had the very much intended consequence of terrorizing the population of Bucharest as a whole.[24] As Romanian officials awaited the next deadly barrage, the Ministry of the Interior excavated the ground beneath major government buildings to create public protection shelters. The ministry, for example, ordered a ditch of about 2.5 meters in depth burrowed beneath the capital's police headquarters along Victoriei Road and then lined it with reinforced concrete. The newly installed air raid shelter, built at a cost of 3 million lei, could accommodate upward of two hundred and fifty citizens.[25]

With the onset of the Cold War, and amidst Romania's push toward a socialist modernity characterized by industrialization and urbanization, underground shelters became a legislated feature of most new constructions.[26] Legislators framed these underground shelters as evidence of the communist government's commitment to defending workers against air attacks and weapons of mass destruction.[27] While rhetorically rich, this material commitment was undermined by the shallow depth of basement-level shelters, their tight quarters, and their general lack of provisioning. The simplicity of the proletariat's shelters also stood in sharp contrast to the lavishly finished and well-provisioned atomic shelters created for political elites. The atomic shelter beneath the Spring Palace (Palatul Primăverii), the primary residence of the then dictator Nicolae Ceaușescu, provides one outlandish case in point.[28]

Figure 30. Movie theater inside the Spring Palace's atomic shelter, 2018. Photo by author.

Now a museum, a tour guide noted that "the palace feels so much bigger in-side than it looks on the outside for a reason: the majority of its rooms are underground, inside the palace's four basement levels."[29] Designed to govern a post-apocalyptic Romania, Ceaușescu's atomic bunker encompasses the Pal-ace's negative third and fourth floors. The aesthetics and comfort of the palace above extend down into the atomic shelter below. The shelter contains confer-ence rooms, bedrooms, and a movie theater. Each are comfortably furnished and elaborately embellished. The shelter's subterranean opulence, as the same guide put it, "helped to hide the differences between the lifestyle that the Ceaușescu family enjoyed and what the Romanian people endured." Pointing to the plush carpeting in the bunker's movie theater, he added, "As you see, during communism, some were more equal than others."

After the Cold War, and as the threat of nuclear annihilation dissipated, authorities did not decommission Bucharest's existing shelters, nor did the production of new shelters cease. Romanian law continued to require build-ings larger than 150 square meters to include a civil protection shelter in their bottom basement level.[30] The issuance of permits for a new building even hinged on the approval of its civil protection shelter's technical design.[31] Once constructed, the law also required emergency services to periodically inspect the upkeep of each civil protection shelter. In this post-socialist moment of privatization, however, responsibility for building and maintaining civil pro-tection shelters, much like the apartment units located above them, transi-tioned from the state to the private citizens who lived and worked within the building.[32] Authorities, meanwhile, insisted that civil protection shelters re-mained relevant, despite having been designed for the previous century's wars. The shelters protected citizens (at least the ones who could afford them) against an increasingly precarious world punctuated by the threat not only of war but also of environmental disaster, terrorism, infrastructural failure, and social unrest.[33] Bucharest's semi-privatized shelter system now sits wait-ing beneath the city to guard against a proliferating list of dystopian threats. Despite evidence of the underground's dangers (see Chapters 4 and 7), au-thorities continue to imagine the underground as a site of security from the generalizable anxieties of the times. Moreover, alongside underground park-ing (see Chapter 5), shelters provide imagined, and perhaps one day mate-rial, security without taking up economically valuable space atop of the city. "We have a shelter because the building permit required it," mentioned one facilities manager for a newly constructed office tower in central Bucharest.

"It's definitely ready for use, if ever necessary. But for now, we don't use it in any way," the manager offered. "It just sits empty."[34]

On the Uses of Shelters

Those caught up in the design, production, and inspection of civil protection shelters openly suspected that their practical utility may very well be imagined rather than real. "Historically, they are for aerial bombings, but now they're for any catastrophe that would cause a building collapse," began one civil engineer who designs civil protection shelters for office towers after I asked when his shelters might be activated. "The thing is," he continued with an uptick of curiosity in his voice, "new buildings are designed to not collapse, even during a violent earthquake."

"I know they're for civil protection, but I'm not really sure what risks they would be useful protecting against," admitted another civil engineer. He continued candidly, "I don't know too much about weapons, but these days I don't think you can hide from them in a basement, even a fortified one." "I could see how they might be useful in an earthquake," offered one architect, "only you're not supposed to use the stairs once the tremor starts." "The use of these shelters is practically none," Iosif, the shelter inspector, bluntly stated. With a tone of incredulity, he continued, "even if you manage to get underground, there's no water and no food in them. And how would you know when to come up? There would be no broadcast signal. And who do you think is coming to rescue you after the city has been leveled?"

For these reasons, among others, the continued production of civil protection shelters in Bucharest stands at odds with the practices and sensibilities of much of Europe. From England to Russia, lawmakers have not just stopped producing new public shelters but have even started to decommission many existing bunkers. Instead of a vital infrastructure with continued utility, as they were considered during the Cold War, subterranean bunkers are now viewed by governments elsewhere in Europe as unremovable relics of a bygone conflict. Constructed to withstand sustained assault by high explosives (if not nuclear blasts), Europe's subterranean bomb shelters continue to endure beyond their perceived utility because their demolition is practically impossible.[35] Officials in Switzerland, for example, estimated the cost of closing its shelter system—much of which is embedded beneath the Alps— at more than $1 billion, an amount that far surpasses the millions needed to maintain those shelters annually.[36] Rather than being demolished and cleared, many of these onetime bomb shelters have instead been creatively repurposed. Given the spatial logic of public shelters, which were placed so as to be accessible from important centers of industry and culture, bunkers often have prime locations beneath major cities that leave them well positioned for commercialization.[37] In this regard, cities across Europe have aesthetically repurposed their bunkers for the enjoyment of the middle and upper classes. Fantastic examples abound.

Take Bunker 42 beneath central Moscow. Completed in 1956 as a Soviet communication facility, Bunker 42 is situated 65 meters belowground and measures over 650 square meters. Designed to withstand a thermonuclear attack, the bunker was equipped with the food, water, and air filtration systems needed to ensure the survival of its 2,500 employees for upward of ninety days. Decommissioned in 2006, private developers transformed the bunker into a museum-cum-entertainment complex, its communication monitoring rooms now a playground for tourists and wealthy Muscovites.[38] Bunker 42's impressive chambers, encased with thick steel-plated ribs and rivets, support the rigorous segmentation of space and the development of simultaneous entertainment experiences (see Chapter 4). Part of the bunker has been converted into a Cold War museum. After paying a $40 entrance fee, the museum invites adult visitors to dispatch long-range bombers bearing atomic weapons, while in a parallel space, their children are equipped with paintball and laser tag rifles to battle the zombie survivors of a nuclear holocaust.[39] In the late night hours, the bunker also contains a concert venue for heavy metal bands, a karaoke bar, and an upmarket banquet hall that can be rented out for weddings and private parties.[40] Tickets to join one luxury party commemorating the end of the Mayan calendar reached upward of $1,000.[41]

In Switzerland, former nuclear bunkers have been appropriated for a wide range of ends, particularly by the hotel industry. La Claustra is one widely remarked upon case in point.[42] Located within a two-hour drive of Zurich and Milan, the hotel took shape inside a decommissioned military bunker buried deep within St. Gotthard Mountain. At La Claustra, the bunker's military-grade blast doors quickly give way to modernist design and comfort. Bedrooms made of glass and steel occupy only a part of the bunker's volume, allowing guests to look out from their rooms onto a softly lit cavern. The seventeen rooms are thoughtfully outfitted with upscale amenities (high-end bathroom finishes and jacuzzi baths), and the hotel also houses a subterranean restaurant. Windows look out onto the cavern's walls, which have been carefully staged with accent lighting.[43] While the aspirational middle classes in Bucharest head underground, sacrificing sunlight (see Chapter 3) and cell coverage (see Chapter 8) to get a toehold in the right part of town, Europe's well-heeled happily pay a premium to disconnect and disappear belowground. "Today wealthy people are willing to pay a lot to get together in the mountain where there is only rock, light, and water and no cell phone reception," the hotel's website boasts.[44] La Claustra positions itself as an upscale retreat from the distractions of modern life as much as a launching pad for exploring mountain trails.[45]

Perhaps most strikingly, the tech industry has taken up and retrofitted nuclear bunkers to protect data instead of people. For all of the rhetoric about digitization and the so-called "cloud," data courses through a material network of underground cables and is increasingly stored on servers located in subterranean bunkers.[46] Built to withstand nuclear blasts, these bunkers purport to protect the data of multinational corporations and billionaire clients alike from the same faceless threats that sustain Bucharest's continued production of civil protection shelters: terrorism, environmental disaster, as well as political and economic unrest. Cyberfort Group, for example, promotes its data center The Bunker as the U.K.'s most secure facility.[47] The Bunker is a converted nuclear shelter in Kent positioned some sixty meters belowground. Its walls are made of three-meter-thick concrete and its entryway is sealed by military-grade blast doors.[48] In the same vein, the Mount10 Company converted a former Swiss military bunker embedded deep within the Alps into what it describes as the "Swiss Fort Knox": the most secure and secretive storehouse for digital information in the world.[49] Mount10 boasts blue chip corporations such as Cisco Systems, Novartis, UBS, and Deutsche Bank among its clients and was entrusted with storing a sealed metal container containing the tools to decipher every major file format then in existence—a kind of Rosetta Stone for the digital age.[50] Beyond Europe, the United States Library of Congress has also adapted a massive bunker in Virginia and filled it with all of the library's movie, television, and sound collections in its effort to prevent a catastrophic loss of human knowledge.[51]

While elsewhere in the world bunkers are being decommissioned and fantastically reimagined to a variety of cultural, educational, and commercial ends, Bucharest's bunkers are not treated as relics of twentieth-century conflict. The Romanian government has very much kept its civil protection system in the present tense. The shelters have withstood the wave of aesthetic repurposing that has otherwise overtaken the Bucharest underground. As the city's bunkers sit in waiting for the catastrophe that might bring them into utility, they actively demarcate in concrete terms the biopolitical contours of belonging in post-socialist Bucharest.[52]

Unequal Inclusion

Should the unthinkable happen, Bucharest's residents will confront a decidedly uneven infrastructure for civil protection. Again, there is not enough shelter space for all; over half of the city's residents do not have a space inside the capital's system of bunkers. The degree of protection that any one citizen will experience in the face of disaster will hinge, in large part, upon their relationship to capital. While civil protection shelters can be found in every sector, they are overwhelmingly concentrated along a band running from east to west through the wealthier center of the city and extending into the affluent north.[53] This distribution reflects the construction boom in office and residential towers, which by law are required to be equipped with a bunker. Simply put, higher earners enjoy a higher density of shelter access, both at home and while at the office, hospital, or government building. Conversely, those living and working in the capital's poorer industrial south and southwest are the most exposed to risk, the most likely lives to be rendered bare in the face of a catastrophe.[54] The residential and employment statuses of the homeless, as the most extreme case in point, preclude them from accessing a shelter through either home or work. In this regard, Bucharest's active network of bunkers serves the same higher-earning populations as the aesthetically repurposed bunkers elsewhere in Europe described previously.

Those fortunate enough to live and work in buildings equipped with civil protection shelters are not necessarily shielded from the world's precarities. By all accounts, the material quality of Bucharest's civil protection shelters reflects the capacity of the populations that they serve to maintain them. As a result, shelter conditions vary widely, from the meticulously maintained bunkers of new office and residential towers to the profound disrepair and dysfunction associated with older, socialist-era housing blocks. Rather than a concrete statement of horizontal comradery, civil protection shelters in post-socialist Bucharest protect the wealthy while rendering the very poor disposable; and they extend an uneven form of inclusion to the middle classes, one in which security and protection has a graduated character.[55]

Unsurprisingly, the civil protection shelters found inside newly constructed office and residential towers offer the highest quality shelter space
for their upmarket inhabitants. These shelters are public only insofar as they
were designed to contain the immediate population of the buildings above
them: they do not accommodate the broader public of the neighborhood or
sector. In an emergency, as the building's occupants make their way downward to the very bottom level of their tower's basement, they will find shelters that are newly constructed and maintained fully to code: dry-absorbent
porta-potties line their walls; the interiors are not only clear of clutter but
swept clean; the lighting is bright and the filters regularly refreshed. As architects and civil engineers insisted to me time and again, the foreign investors who own and operate new residential and office towers are keen to prevent
legal requirements from becoming liabilities. They therefore tasked their
facility managers with making sure that their building's civil protection
shelters are ready for inspection (even if not for deployment). "It's always
ready should it be needed—we even give it a thorough scrubbing every
spring," a facilities manager for SkyTower explained. "Otherwise, we don't
use it. We already have all of the parking and storage that our tenants could
need," she said with a shrug. Alongside the requisite sanitation equipment,
another office tower in the city center has added to the practical utility of its
civil protection shelter by affixing seismic sensors to its walls. The sensors'
measurements, the inspector explained, provide insight into how the building would endure through a high magnitude earthquake. "We can check if
the building performed as expected, and that'll let us know if the buildings
is still safe for use."

For oligarchs who prefer spacious villas to penthouses in newly constructed towers, companies such as Atomic Bunker Romania offer a premium, private bunker experience should catastrophe strike while out of the
office.[56] Starting at €130,000, the short-lived Atomic Bunker Romania installed their reinforced concrete enclosures at least six meters belowground,
beneath the family villa. As with other civil protection shelters, Atomic Bunkers were designed to withstand a world gone sideways in the wake of a wide
range of events: earthquakes, nuclear and chemical accidents, terrorism, or
warfare, as well as the chaos that might follow infrastructural failures such
as sustained power outages.[57] Unlike civil protection shelters, however,
Atomic Bunker Romania couched its bunkers in terms of intimacy and inhabitability. These bunkers were not public in any sense: they only accommodated the nuclear family, and so Atomic Bunker staged its shelters to affect

a sense of home. Their internal organization featured a living room with a television, dining, kitchen, and sleeping areas, and a bathroom with water and sewage treatment. These bunkers were not as highly aestheticized as those that have been repurposed as luxury hotels or nightclubs elsewhere in Europe; Atomic Bunker Romania was not selling leisure spaces to the wealthy. They developed bunkers as active civil protection shelters where clients could retreat when they needed to (rather than as leisure and culture sites they could visit when they wanted to). Nevertheless, should the time of need ever arise, the company's clients would find in their bunkers a warm atmosphere with modern finishes where a large family, the company assured, could survive for a week without electricity, making these private bunkers not only more comfortable but also more useful than public shelters.[58] Such bunkers equipped the families of the fortunate with the means to endure through the apocalypse, even as the rest of society did not. For those further down-market, Atomic Bunker Romania offered to transform existing cellars and basements into emergency shelters equipped with air filtration and purification systems for €2,500, extending to Romania's suburban homeowner an American fantasy of survival.[59]

The individualized sense of domestic comfort evoked by Atomic Bunker Romania's private household shelters stands in stark contrast to the majority of civil protection shelters found beneath Bucharest's housing blocks and governmental buildings. Built over the course of the twentieth century, these public shelters reflect the same design criteria as those found in new office and residential towers. However, the older shelters serve middle classes with a varying capacity to maintain them properly. "The simple fact is that there aren't that many inspectors," Iosif explained. "There's enough staff to check paperwork on shelters but not enough to make regular site visits," he continued. In the absence of government oversight or support, the state has left housing block residents with their own tight budgets to agree to share the cost of the bunkers' upkeep. "The system assumes a block filled with residents who can afford to contribute. That's hardly ever the case," Iosif observed. "All the time I heard people ask, 'why should I pay for a collective shelter when others don't?'" Placing the onus for shelter upkeep upon Bucharest's middle classes forces the underpaid and overworked to direct portions of their limited household budgets toward a shared resource of questionable utility. Not surprisingly, many of the city's residents shrug at this civic obligation and instead focus their resources upon maintaining and improving their home life, sending Bucharest's shelters into various states of disrepair.[60] Citing sources from City Hall, one columnist estimated that 50 percent of civil protection shelters are flooded, recasting Bucharest's archipelago of underground bunkers from impractical to unusable.[61] This point resonated with Iosif. "Even if you fix the flooding, the filtration systems will be rusted out. And if you fix the filtration, and the shelters remain dry, the building's residents typically put them to use as extra storage space. The dry shelters beneath housing blocks are too crammed with stuff to be accessible in an emergency," he insisted.

The heightened inequality within the capital's civil protection shelters reflects, ultimately, a degraded sense of belonging at the very foundation of Bucharest. Where once the government could reasonably ask its citizens to rely upon each other to persevere through a common threat, today's city residents must self-finance their security. In response to such a governmental obligation, market differentiation has concretely entered into public life in ways that nuance the politics of inclusion and exclusion that dominate so much thought about what it means to belong to the city. While the lives of the excluded are no doubt rendered bare, those included within the city's civil protection system are not necessarily protected. Rather, the aged residential blocks marketed to the middle classes offer only an uneven form of belonging. These shelters afford inclusion at the level of extending to the middle classes the financial obligation to maintain a shelter, but such shelters are of such a degraded state as to offer little (if any) protection from real danger. Such inequalities of inclusion chart in concrete terms the biopolitical limits of the city—a city in which the middle classes live and work in an existential state of semi-disposability.

CHAPTER 10

Tombs

As Bucharest developed ever further downward, no underground real estate market heated up faster than the market for burial plots within Bucharest's public cemeteries. In 2012, Bucharest's news outlets reported with bewilderment that plots inside the city's oldest public cemeteries, which are distributed by the Cemetery Administration (Administraţia Cimitirelor şi Crematoriilor) for roughly €60 in fees, were being resold online for between €5,000 and €10,000.[1] The gray market for burial plots inside Bucharest's most venerable public cemetery, Bellu Cemetery, was especially robust. Tombs inside of Bellu were being sold for over €20,000—rivaling the cost of a studio apartment.[2] Such sums were all the more extraordinary given that no one goes unburied in Bucharest: the city guarantees a plot, at no cost beyond the nominal administrative fee, to anybody who requests one. Moreover, even though fourteen of the municipality's fifteen public cemeteries are officially full, the Romanian Orthodox Church permits the reuse of burial plots.[3] The vast majority of the city's residents already have established family plots that can be reused in times of need. Administrators were also quick to note that ample room remained inside the city's newest cemetery, Străuleşti II, to accommodate the city's newest and least-established residents.[4] What the Cemetery Administration could not accommodate, given its occupancy rate, were requests by the metropolis's social climbers to replicate their trajectory in the necropolis, relocating from the likes of Străuleşti to Bellu, for example. And so, at the dawn of the twenty-first century, there was no shortage of Bucharestians willing to create a new market around a public good to meet their desires, with some paying an extreme premium to cement their newfound status in the hereafter, and others ready to cash in on that inclination.

Figure 31. Tomb, Bellu Cemetery, 2019. Photo by author.

While Bucharest's upwardly mobile retreated behind the privacy of gates, walls, or penthouses in life, the grey market for public cemetery plots evidenced the opposite desire in death.[5] As any upmarket funerary director will explain, Bucharest's most ambitious and status-conscious eschew the capital's bourgeoning private cemeteries, with their own promises of distinction substantiated by luxury finishes, to instead secure a plot inside Bellu: the city's oldest, most centrally located, and most heavily trafficked public cemetery.[6] Unlike in the metropolis, where the underground is usually regarded as less desirable than the city surface, the underground of the necropolis is the most sought-after kind of space—and nowhere more so than in Bellu. As the city's first cemetery, Bellu is where the capital's founding families have maintained their plots for generations. Unlike any other cemetery in the city, Bellu offers the upwardly mobile the opportunity to take their place alongside not just the capital's but also the nation's most prominent contributors to the arts, culture, science, politics, and industry.[7] The distinction also rubs off on their descendants, who in their mourning might brush shoulders with Bucharest's most prominent households. Try as private cemeteries might to build a sense of cultural cachet, their fine finishes and tasteful landscaping cannot compete with history and precedence.

This book's final chapter turns toward Bucharest's public cemeteries to analyze the impulse of the upwardly mobile to spend significant sums of money on burial tombs. It moves between the extreme poles of the Străuleşti and Bellu cemeteries, which serve the city's least- and best-established families, respectively. Buying into the right burial plot, this chapter argues, holds the promise of escape from the many degradations of a middle-class life, with its second-tier claims to status, encapsulating the dream of an uncompromised belonging for all of eternity. Perhaps paradoxically, citizens' sense of belonging within a national public is nowhere more durable in Bucharest than in death.

Bellu Cemetery

"I have been informed that the dead in the Capital are still being buried in church grounds because the land needed for cemeteries has not yet been allocated," Baron Barbu Bellu wrote in 1855. "In this regard, I made a proper agreement with the Minister to cede to the City Hall the estate needed to set up a cemetery."[8] A former Minister of both Culture and Justice as well as a

Figure 32. Bellu alleyway, 2016. Photo by author.

onetime judge at the Illov County Courthouse, Baron Bellu's gift of his garden on Şerban Vodă Street reflected a broader modernization movement
among civic leaders sweeping across Paris, London, and New York driven by
the emergent fields of public health and urban planning.[9] In Bucharest as well,
moving death and decomposition from parochial cemeteries sprinkled
throughout the city and containing it in newly created municipal cemeteries
on the outskirts of town became, at that moment, a necessary strategy for
controlling disease within the city.[10] Unlike the targeted approach that would
eventually guide the development of civil protection shelters (see Chapter 9),
the perceived threat that any dead body posed to public health drove the inclusion of every dead body within the city's municipal cemeteries.[11] To that
end, the Şerban Vodă Cemetery, or Bellu Cemetery as it has come to be
known, opened in 1858. To encourage its use, Bucharest's founding families
moved their ancestors' remains to Bellu shortly thereafter.[12]

As Bucharest's public cemeteries opened their plots to all, they became a
microcosm of the city's inequalities and contradictions.[13] In the decades that
followed, Bellu Cemetery's growth reflected Bucharest's own flourishing. The
cemetery became the final resting place of the civic leaders and captains of
industry who proudly oversaw Bucharest's emergence by the early twentieth

century as a center of European commerce and trade.[14] Bucharest's founding fathers used their growing wealth to commission leading Romanian and
foreign sculptors and architects to build grand tombs and monumental statues to commemorate their family burial plots.[15] By the mid-twentieth century,
Bellu Cemetery had become widely regarded among scholars, politicians, and
residents alike as an open-air museum as much as a cemetery.[16] Alongside
Bucharest's founding fathers, and beneath its great works of art, Romania's
most notable contributors to knowledge and culture—such as the poet Mihai Eminescu, the philosophers Emil Cioran and Mircea Eliade, and the play-
wright Eugene Ionescu—took their place inside Bellu Cemetery. Despite its
unmistakably bourgeois history and aesthetics, Bellu remained the preferred
burial place for elites during the socialist era, attesting to the enduring cultural value of its plots.[17] By the twentieth century's end, Bellu Cemetery had
reached its full capacity. While administrators insisted to me that Bellu was
an inclusive cemetery that was always available to all, the consistent burials
there of leading figures, situated beneath elaborate statuary, suggests a deliberate project of national commemoration and image-making. The regular
descriptions of Bellu in scholarly and popular writing as a place commemo-
rating the "nation's pantheon" of historical figures, similarly suggest as
much.[18] Regardless of one's contributions to the city's civic, financial, or cultural life today, though, the Cemetery Administration has no more available
plots to dispense to those families who have not already been admitted.

Although officially closed, Bellu Cemetery very much remains active;
burials occur there regularly. This is because the Romanian Orthodox Church
permits the reuse of burial plots, making space inside Bellu Cemetery an
important family resource. As Father Simu, an Orthodox priest stationed at
the Patriarchal Cathedral, explained to me, an elaborate cult of the dead
organizes the process of burial, commemoration, and, ultimately, the freeing of a plot for the next family member. The cult of the dead is punctuated
by the performance of several religious services for the funeral itself and then
for regularized memorials for seven years thereafter.[19] These ceremonies are
both highly detailed and gendered, requiring women to master complex rituals over the course of their lifetimes.[20] The ceremonies are also intensely
social in character, calling to the grave with some regularity the bereaved's
family, friends, and neighbors. After the seventh year, the deceased's bones
can be ceremonially collected and moved to the foot of the grave, readying
the plot for its next occupant. The reusability of these precious plots sets the
condition of possibility for their reselling.

"The cemetery must also be cared for—you cannot leave a plot in a state of decay," Father Simu warned. This point is critical for understanding the supply of plots in Bellu. "Plots come with obligations: they have to be permanently cared for even if there is no one resting in it." Care involves both the paying of nominal annual taxes and fees to the cemetery and more laborious maintenance of the plots themselves: keeping the grass trimmed and headstones scrubbed, planting or leaving flowers. Citizenship in death is not without responsibilities to be performed by the living. As Gail Kligman observed, the cycle of death-related practices culturally encodes norms of social behavior that associate the self and society as well as the living and the dead so that the dead may be out of sight, but never completely out of mind.[21] When properly maintained, the plots stay within the family. The failure of the living to care adequately for their ancestors' graves results in the plot being labeled "abandoned" and being made available for reallocation by the Cemetery Administration to another family. "They're kind of like homes," as one funerary director in Bucharest put it to me. "As long as you pay your taxes and keep it in good condition, it's your place. You can pass it down." If not, they get repossessed and reallocated.

The demand for these burial plots is clear enough. Bellu offers a sense of legacy for the deceased. Given the obligations for living family members not only to regularly visit and maintain their plots but also to host family and friends at them, the placement of burial sites in Bucharest has status implications for the living. "Everybody wants to have a plot closer to their home," Father Simu explained. "You want the cemetery to be ten minutes, maybe a half hour, away. That's the ideal. You don't want it to be an hour and half or two hours away." Every Romanian of a certain age I spoke with said as much. Accessible plots are desirable because they facilitate the routine payment of respect to the dead by easing the burden of time and travel placed upon the living. And no cemetery in Bucharest is more central than Bellu. It can be reached within five minutes from Union Square by car or Metro. Internally, Bellu Cemetery is organized around a system of drivable roads and alleys. Sitting on a bench in Bellu, it is not uncommon to see mourners delivered to the foot of their tomb by taxi. As one might expect, the very opposite is true of Bucharest's only open municipal cemetery, Străulești II. Located along the city limits, it takes around ninety minutes from the city center to reach Străulești II by bus or car, depending on traffic.[22] Distant plots quickly raise uncomfortable speculations among older Romanians about the will and capacity of their family and friends to properly remember them.

The desirability of a burial plot hinges, ultimately, upon both its location in space and its perceived rank in a social order.[23] "There are three important cemeteries if you are from something like a prominent family in Bucharest," Denis, a high-end funeral director, explained to me in 2018. "There is Bellu and then there are also Ghencea Militar and Sf. Vineri Cemeteries." These are three of Bucharest's oldest cemeteries and their prominence is tied as much to the families they serve as to the elaborate works of art that make up the tombs and mausoleums found within them. "There are preferences by neighborhood," Denis insisted. "It's clear that people who live in Ghencea prefer Ghencea Militar. They wouldn't want to be in Sf. Vineri because it's harder to get there." While most residents are more concerned with proximity, Denis explained, those with means are interested in moving their families into cemeteries with status. "Bellu is the most historic, it is the ultimate, the pinnacle, place to be."

If visitation is prized in death, no cemetery in Bucharest is higher trafficked than Bellu. Since EU accession, officials at City Hall have worked to position Bellu as a major tourist destination of European significance. In 2010, for example, the Cemetery Administration successfully petitioned for Bellu Cemetery to join the Association of Significant Cemeteries in Europe (ASCE).[24] The designation placed Bellu alongside other storied national cemeteries such as Père Lachaise in Paris and Highgate in London. A commemorative volume of photography, assembled in the style of a glossy large-format coffee table book and produced by City Hall, affirmed as much.[25] By adding Bellu to the list, officials hoped to tap into heritage tourism circuits that attract upwards of five million people annually.[26] At the same time, Bucharest's City Hall received an €800,000 grant to officially transform Bellu Cemetery into an open-air museum.[27] Bellu quickly became one of the most popular museum destinations in the capital, drawing no fewer than twenty-five thousand visitors during the city's annual "night at the museum" festival in 2015.[28] Funerary companies serving the upmarket neighborhood of Dorobanți insisted that these designations only add to the sense of legacy that comes with a family plot in Bellu.

The demand for a burial place in the national pantheon has driven Bellu to maximum capacity, leading to problems of density and mobility that mirror those of the city as a whole. "Have you been behind its chapel?" asked Viorel, a funerary director serving clientele in Bucharest's center-north. "They [the Cemetery Administration] have used every square centimeter. You can't take half a step without walking on top of someone's grave. There's no pathway for visitors—they have to jump over headstones and around tombs to visit a grave. Some of those bodies must be stuffed under the chapel in order to fit in. There's nothing nice about it," he chided. "But my clients will happily pay €5,000 for one of those spots, just to be in Bellu." The basic approach to internment extended throughout Bellu, with plots compactly placed cheek by jowl leaving little room for the living to navigate the cemetery without inadvertently stumbling onto the memorials of others. And there is no possibility for Bellu Cemetery to expand: the city of Bucharest has developed around Bellu, and other municipal cemeteries, in ways that fix their borders.[29] "Especially after 1990," Father Simu explained, "every space in the city has been absorbed by continuous construction." Pointing to the Liberty Center Mall, which opened near Bellu in 2008, he continued, "In the context of Liberty, even if there was open space, the land would now be too costly to use for burials."

Poorly Placed

Despite the steady drumbeat of noisy headlines about overfull cemeteries and soaring secondary markets for prestige burial plots, a senior administrator for the Cemetery Administration named Toma was clear on one point. "No one in Bucharest goes unburied," he told me definitively from behind his office desk in 2016. Corpses do not stay stacked at overwhelmed morgues; they are not cremated for lack of room, nor are they dumped in communal ossuaries. The municipality affords each and every resident of Bucharest the right to, as Michel Foucault irreverently put it, "his own little box for her or his own little personal decay."[30] Toma explained, "We continue to dispense places of internment. If someone needs a plot, a relative can come with a death certificate, and we will give them one on the spot for a term of at least seven years." The municipality's unwavering commitment to burial marks Bucharest's public cemeteries as important sites for providing belonging, even to the city's most vulnerable. Officials make no guarantee, however, that everyone will be honored equally. Some are highly commemorated while others are not at all.[31] As Toma added earnestly, "The available plot will be at the margins—in Străulești—but it's a plot nevertheless." Viorel, the upmarket funerary director, affirmed Toma's characterization of Străulești as marginal, in cultural as well as spatial terms. "It is definitely a marginal cemetery—there is no one famous or historically significant buried there. It's a cemetery for

Figure 33. Străulești Cemetery, 2016. Photo by author.

the very poor, or for villagers living on the outskirts of Bucharest. It's not for people with means." While an undisputed site of incorporation into the necropolis, the kind of citizenship on offer at Străulești is, from a certain middle-class perspective, a degraded one. If internment at Bellu cements one's legacy in "Romania's pantheon" for eternity, a burial at Străulești places one at the periphery of the city and its social order.

Though snobbish in his tone, Viorel is not out of step in his assessment of the order of things. The Cemetery Administration confirmed as much. Sitting in his office, Toma spoke frankly: "Everyone wants to be in Bellu, but it's a closed cemetery. The only funerals happening in Bellu are for people with family plots, inherited from father to son from the 1850s until today." In Toma's view, a plot in Bellu is more of an inherited resource for Bucharest's old families than an earned honorific. "But Străulești is for the new arrivals, the poor, the unidentified," he continued. "Those are the people who go to Străulești. It is the one cemetery where we still have availability, and it allows us to offer everyone a dignified burial there." The Cemetery Administration's admirable and unwavering democratic commitment to a universal belonging in death was put on global display when the British television program *Channel 4 News* in 2014 aired a string of segments about a homeless encampment that had formed in the sewer canals beneath Gara de Nord Station (see Preface). Straining the ethical limits of representation as it trafficked in sensational images of addiction and immiseration, *Channel 4 News*'s story focused upon the death of an unhoused teenager from complications related to AIDS. With references to drug consumption sprinkled throughout, the segment followed the teenager's funerary procession from the sewer canal where she had lived to the gravesite at Străulești where she was ultimately buried and mourned by those also living in the sewer.

Such burials are inspired statements about the equality of citizens in the face of death. For Viorel, however, such moments also represent the kind of spectacle of poverty that his upmarket clientele wish to avoid for their families during their own time of mourning. To be sure, tensions surrounding class and ethnicity run through the necropolis as much as they do the metropolis. The upwardly mobile in particular are anxious to demonstrate their own respectability while at the same time avoiding proximity to those they perceive as racially or economically other. The tension revolves around questions about the presentability of the dead's mourners—with concerns over the unhoused person's soiled clothes and soft drugs, for example—and also extends into the judgement and taste shaping how the graves themselves are materially staged. "It's not that everyone at Străulești is poor," Viorel continued; "there are some massive palaces on display there. And they're very expensive constructions too. But they're clearly built by gypsies [*ţigani*]." Given the baroque aesthetic of Bellu, Viorel's scorn for these graves was evidently based on judgements about taste and excess rather than a particular artistic style.[32] He struggled to contain his laughter as he described the cemetery's more elaborate memorials: "In total seriousness—you'll find memorials there that look like small homes built of black marble and gold, with thermopane doors and even a water feature," he cackled. Viorel took a moment to collect his breath before adding, "No Romanian would build something like that." Sandu, another funerary director working in the wealthy center-north offered a more tempered assessment, albeit in similarly racialized terms. "The Roma are building garish mausoleums at Străulești. Romanians just don't build like gypsies. Go have a look for yourself—when you see a massive and ugly structure, it's probably a gypsy's monument."

While providing incorporation, Viorel, Denis, Sandu, and other funerary directors in Bucharest agreed that Străulești does not offer respectability, nor could it. The particular problem confronting Bucharest's upwardly mobile was that no sum of money, however tastefully spent, could overcome the low social standing of such a cemetery. "Everyone knows about Străuleşti, because that is where the state is handing out plots," Denis explained. "If you're new to Bucharest, it's where you get placed. And if your family has some potential, you're not spending money to try and make Străuleşti nice. That makes no sense," he declared definitively. "Instead," he continued, "you'll spend the money to get into a cemetery with some renown." As one might expect, those who can't afford to spend the money don't. "The only people being buried in Străuleşti are the ones who don't have a car, or they just don't know better," Viorel observed. His description of Străuleşti's residents closely mirrors accounts of the city's Metro riders as those who are too poor to drive (see Chapter 2). To avoid being consigned to a déclassé location for all eternity, the upwardly mobile provide the demand needed to sustain a semi-private market for desirable public burial plots. As the status-conscious shell out large sums to move toward Bellu and away from cemeteries like Străuleşti, they deny the equality of death implied by the Cemetery Administration's inclusiveness. Instead, the necropolis has emerged as a key domain for establishing social inequality in the afterlife.

The semi-private market for plots in prestige public cemeteries was separate from the emergence of private cemeteries. The plot shortage in desirable public cemeteries has spurred the growth of a string of private cemeteries on the outskirts of Bucharest.[33] While municipal rates vary between €1.50 and €55 for a burial plot, private cemeteries charged between €1,650 and €3,500, with the total cost of a funeral easily reaching €10,000.[34] The sheer cost of internment places private cemeteries beyond the reach of the lower classes. Their administrators also enforce a certain aesthetic within the cemetery: plots are spaced evenly while headstones typically require approval if not purchased directly through the cemetery.[35] While more respectable than a burial in Străuleşti, the families I spoke with who had bought into a private cemetery experienced the burial of their loved one as nothing short of a shakedown. "The priest told me the headstone I bought through the cemetery was not in compliance with its own standards and needed to be rebought," explained one man. "My family was asked about package upgrades during the service," reported another woman. A third remarked with disbelief that she was charged for weeding and grass clipping after having paid several thousand euros for her family's plot. Alongside its worries about the unseemly mix of death and profit maximization, the Romanian Orthodox Church voiced its own concern about newly formed corporations overseeing eternity. "What happens if, after a number of years, the firm goes bankrupt?" the representative from the Patriarchy, Father Simu, asked rhetorically. "This is why the church and the state ought to be administering cemeteries. They don't disappear as easily," he reasoned.

While private cemeteries offer exclusivity, in their nascent form they lack not only polish but also distinction. Their clients trade a degraded sense of belonging within the marginal public cemetery for consumer discontent in a private one. Either way, an aura of compromise looms in the background. This is why Romanians with means have turned to grey markets to buy into prestigious municipal cemeteries, and particularly into Bellu. There, in the soil of Bellu, full belonging can be attained without any hint of compromise. The desire to buy into the right place fundamentally reframed the economy of death. "With my clients," Denis observed, "the most money is spent at the cemetery, for the burial plot. Hardly anyone is spending money on the quality of services." This trend stands in stark contrast to the sensibilities observed just a few decades ago, in the socialist era, when families spent sums that rivaled the cost of a wedding upon the funerary service itself as well as on the meals for the obligatory days of mourning that followed.[36] Burial plots, as a publicly provided good, were not a major expense. Today the opposite is true. Viorel, in an effort to contextualize the changing sensibilities of Bucharest's up and coming, put it to me this way: "To be clear, no one is paying hand over fist for a simple plot. They're buying a legacy. And in today's Romania you can buy whatever you want. Why should this be any different?" Cracking a smile, Viorel added, "I've had plenty of pensioners making €200 or €300 a month coming here, ready to pay out the money for their legacy."

Moving on Up

The market for burial plots in Bellu, and other prestigious municipal cemeteries, is as grey as it is robust. This is because Bellu's burial plots are officially public property. Even though certain burial plots have been tied to prominent families for generations, their plots are not their family's property. Rather, people receive a concession to use a burial plot, and so long as the site is properly maintained and the appropriate taxes paid, they not only keep their concession but can pass it down generationally. "Naturally you cannot sell something that is not your own," Toma from the Cemetery Administration explained to me, "and this includes cemetery plots. But the concessioner is allowed to donate their plot to someone in need. These donations used to be benevolent—like giving the plot over to a cousin. But now"—here Toma grinned sheepishly—"they're being 'donated' to strangers for exorbitant sums." The lucrative market that has emerged around municipal graves elicited awkward chuckles and uncomfortable shifts in chairs from Cemetery Administrators and Orthodox priests alike. The gentrification of the Bucharest underground has been a consistent project of transforming public spaces into semi-private property for the sake of personal profit, and the personal appropriation of public cemetery plots strikes many as the most brazen case in point. Despite the occasional handwringing, however, these transactions occur both publicly and unapologetically. "It's an abuse of the terms of the concession, but it's not illegal," Father Simu assessed. With a hint of resignation, he added, "It's more of a matter of poor morality."

OLX.ro, a Romanian website similar to eBay, organizes much of the traffic in Bucharest's municipal burial plots. The company justified its involvement in the process in terms of the liberal, moral value inherent in structuring any market: doing so facilitates freedom of choice.[37] "The charm of OLX.ro," a company representative wrote to me, "is that it can sell and buy anything within legal limits, and this is great for everyone." Recasting municipal burial plots as a "form of property," the representative reported several hundred transactions for burial plots each month on its site, with buyers generally paying the listed price.[38] The listings themselves were remarkably flat. They typically offered three descriptive sentences juxtaposed with startling asking prices. "It is built like a church, imposing and tall, with interior painting," read one listing for a €40,000 tomb. Information that one might expect when acquiring a piece of art, such as the architect's name or the date of the work's completion, for example, was not offered. A string of images, however, sought to confirm the tomb's grandiosity of presence. Another listing for €28,000 boasted of the tomb's close proximity to the Raffaello Romanelli sculpture "Woman with Umbrella" ("Doamna cu umbrella"). Half of this listing's images showed the "Woman with Umbrella" from the perspective of the tomb's foot, helping to structure the griever's visual expectations. For a few thousand euros, the least expensive listings in Bellu mentioned very little about the plots themselves and instead drew attention to the history and atmosphere of the cemetery at large. "Keep in mind that a typical funeral in Bucharest, with everything included, is about €1,300," Denis contextualized for me. "Even with the least expensive plot in Bellu, we are still discussing extraordinary prices." With a shrug he added, "There are plenty of people who pay for it."

While the upwardly mobile maneuver to buy the right plot in the necropolis, the downwardly mobile make their plots available. "My family tomb
isn't painted by just anyone," one plot holder told me. "It's a work by Nicolae
Grigorescu."[39] While commentators note that the asking prices for such listings rival those of a studio apartment, those cashing in on their family tombs
worry that the money will not go far enough.[40] Instead of relishing the hand-
some sums they sought to bring in, plot holders spoke of the constantly rising cost of living in Bucharest. Using the language of "selling" rather than
"donating," the same plot holder griped, "If I sell for the asking price, I can
maybe buy a good house in another city in Romania, but not in Bucharest."
"The price of an apartment in Bucharest now costs the same as in New York,"
another plot holder said to me without any hint of hyperbole. Not unlike the
homeowners who have converted their basements into apartment rentals (see
Chapter 3), these plot holders hoped to shore up their lifestyle in an increasingly expensive capital city by commercializing their space underground. A
third plot holder, for his part, had already migrated elsewhere in the EU and
was eager to liquidate what couldn't be carried abroad. Each of these plot
holders reported establishing their asking price in consultation with Bellu's
staff. "Nothing moves at Bellu without the priest involved: they set the price,
they control everything," one plot seller told me. "There is a great mafia inside Bellu," insisted another. "It's very hard to do anything without them."
Such claims resonate with investigative reporting circulating at the time.
News accounts documented private speculators who paid gravediggers tips
to identify Bellu's abandoned plots and then acquire and sell those plots for
between €3,000 and €30,000.[41] Administrators at Bellu Cemetery, for their
part, strongly denied such accusations. "There are rules to keeping these
plots," Toma replied when asked in 2016 about such accusations. "If the burial
plot is not tidy, if taxes haven't been paid for ten years, we can initiate a pro-
cedure and ask the Cemetery Administration to revoke the concession right
because the plot is not being cared for." Without conceding wrongdoing,
Toma continued in practical terms, "And sure, it is the diggers who know
which plots are no longer maintained, which plots are no longer visited."

Beneath the moments of public outrage and bureaucratic denial, the upwardly mobile prove all too willing to help those slipping down the ladder
find their footing, if only for a time. While the private option predominates
in life, the sense of belonging offered by municipal cemeteries remains the
most prized possibility in death. Simply put, the institution of national citizenship offers the departed the most enduring legacy. Against the million mo-

ments of indignation that come with their second-tier status, the middle classes eagerly recast public goods as semi-private luxury commodities to acquire the most compelling form of citizenship in death: burial amongst the national pantheon at Bellu Cemetery. By being interned there, within its storied grounds, the upwardly mobile hope to fully shed their anxieties and quell their grumblings so as to finally rest in peace.

Conclusion

"The most growth in residential projects in Bucharest has been at the periphery, in the deindustrialized zones," Lucian, a civil engineer working for a large developer, explained as we chatted in his corporate office. He switched out a tailored blazer for a field jacket in preparation for our site visit to a project underway in the western suburbs. "Firms are making large profits off of big developments out there because large land plots can still be bought on the cheap." Given the frenetic clip of development, available plots of land within the city center had become hard to come by, and the few that remained were exorbitantly expensive. Changing his designer sneakers for construction boots, Lucian smirked as he continued: "And you should visit these neighborhoods first thing in the morning. The strange thing you'll see is that entire developments are filled with professionals, and they're all fleeing these neighborhoods on the Metro."

And "flee" they do.[1] Moreover, their flight from the periphery is made possible specifically by the underground. When I first visited the deindustrialized zone of Berceni in 2006, it had only recently begun to transition toward residential housing. At that time, factories with billowing smokestacks dominated the landscape while only a handful of suburban homes had taken shape in the distance. Their new owners publicly questioned their purchases to reporters; they described the polluted red smoke pouring out of the factories throughout the day, rendering the air difficult to breathe.[2] Despite these conditions, however, over the next decade Berceni became a hotspot for residential developments, featuring dozens of housing blocks. By 8 A.M. on a workday in 2018, its Metro station was a hive of activity. However, unlike centrally located stations, which have a steady mix of people coming and going, all of the passengers in Berceni were outbound. By 10 A.M., the residential developments clustered around the Metro looked like ghost towns. The sidewalks were empty, and I strolled down the middle of main roads without being disturbed by a car; the convenience stores selling essential groceries, cell phone credit, and coffees to go were entirely without customers. It was

Figure 34. Evening rush hour, Berceni Metro, 2018. Photo by author.

not until the end of the workday, when a heavy flow of Metro passengers emerged out of the underground, that the neighborhood returned to life.

Dan, the owner of Comfort LLC (see Chapter 1), helped to bring this dynamic into clear relief. "In the ultra-center, residential sale prices can reach well over €1,800 per square meter. But at the city limits, in places like Berceni, units go for only €700 per square meter because no one with money would ever want to move there: it's ugly, there's nothing around, the commute is terrible," Dan rattled off. To be fair, the factories that used to billow so much smoke in Berceni then sat idle, but they were falling into a state of disrepair. Along the deindustrialized periphery, autobody shops and defunct factories outnumbered cafes while the boutique shops, slick bars, or fine dining options that typically cater to the middle classes were nowhere to be seen. The well-worn world beyond the walls of these new residential developments stood in stark contrast to the freshly painted condo units outfitted with Ikea-inspired kitchens and marketed to the middle classes. And at least for those travelling by car, Dan was correct to say that the commute was terrible. "But what we found is that the Metro changes the buyer's calculus entirely. If we develop our project within a five-minute walk of a Metro station—even ones at the very end of a line—we can price those units as if they're next to the city center, upward of €1,100 per square meter."

Promotional materials for residential developments along the industrial periphery made crystal clear the transformative presence of the underground. As a genre, these materials framed glossy renderings of new residential

Figure 35. "Only 2 minutes from the Metro," 2017. Photo by author.

complexes with two pieces of information: the project's name in bold font located above the image, and, in a slightly softer font, the approximate walking time to the nearest Metro station. The shorter the walk, the more marketable the development. Palladium Residence, for example, promoted its four-minute walk to Nicolae Teclu Station. Rotar Park Residence, meanwhile, boasted of a two-minute walk to Păcii Station, while Victory Residence in Berceni advertised a mere one-minute walk to Dimitrie Leonida Station. Time and again, developers packaged and sold their buildings along the industrial periphery based on their proximity to the Metro rather than the quality or character of the areas that immediately surround the developments themselves.

As Dan insisted, suburban developments in peripheral zones with working-class connotations and "ugly" aesthetics were pitched to the middle classes by way of the sub-urban. "These zones are traditionally workers' zones—they have a connotation of being for the lower middle classes in terms of income and education," Dan stated as fact. And yet, by positioning his peripheral developments in relation to the underground, Dan successfully sold his units to newly minted professionals. "They don't have extraordinary family resources," he continued, "and it's their first job in Bucharest; their salaries are still relatively low, and they're covering the down payment themselves. They can't afford to buy and then renovate more centrally located properties, and the commute by car from out here would otherwise be too much. The Metro is what makes living here attractive to them."

The opening up of the periphery for middle-class housing developments hinged upon the underground's ability to allow residents of these developments to feel more closely connected to the city center than if they had to inch their way there through traffic. The Metro's relatively compressed travel time enabled residents to imagine themselves not only as being near the city center but also as being fundamentally separate from the deindustrial neighborhoods in which their buildings were actually located. As Gabi, a resident of a Comfort LLC unit in Berceni, insisted to me one evening over beers that he procured from a micro-brewery in the center-north, "The people living inside this place are of another kind than those living just down the road." Gabi's claims to entitlement and status were all the more remarkable given his own provincial background. Originally from Moldova, Gabi moved to Bucharest for university to study architecture. After collecting his degree, he stayed in the capital and started his own firm with his wife. Although the two run their business out of a home office in Berceni, Gabi dressed up in an H&M suit and rode the Metro into the city center to take his client meetings in Old Town's fashionable bars and cafes. He also spent his Friday and Saturday nights in Old Town's underground clubs, partying until the early morning when Metro service resumed to ferry him back to Berceni. Although Gabi lived *in* Berceni, he insisted to me that he was not *of* Berceni. Most every day, Gabi took the underground and was shuttled into the city center, where he understood his dress, education, occupation, and comportment as making him more akin to the professionals found there, miles down the track, as opposed to the working classes living just beyond his development's entryway. Gabi went so far as to claim that those who were *of* Berceni would not fit in in the city center as he and his neighbors from the development did. Those who have bought into developments like the one he lives in, Gabi continued, are: "educated, 'cultured' [*civilizat*]; they ride the Metro into the center and know how to behave when they get there," while those living outside them, in Gabi's estimation, do not. Others I spoke with inside Gabi's building voiced similar sentiments.

Gabi brings into focus the kind of middle-class subject that the aestheticized and expanded underground makes possible. His sense of middle-class respectability, relative to the working classes he physically neighbors, was founded entirely upon the underground. During the workweek, the Metro system allowed Gabi to style himself as a middle-class professional rooted in a city center that he was in fact priced out of; on the weekend, Old Town's basement clubs did the same. The underground also opened up to Gabi the otherwise unthinkable sense of respectability that came with buying into new housing construction with modern finishes. Even if his building was in the wrong part of town, the Metro shuttled him quickly and easily into the city center, where he insisted he truly belonged. By way of the underground, then, Gabi could organize his workweek and his leisure time in the city center, despite being priced out to the city's outskirts. For others, station kiosks, basement conversions into apartments or offices, or even grave plots served a similar function. These underground spaces organized a growing set of claims upon a densely populated and overbuilt city.

In this sense, the efforts of city planners and private developers to order an impossibly congested city by way of the underground have succeeded. New Metro lines, parking garages, as well as basement and cellar conversions all help to expand the city to accommodate new middle classes. In this transformation, the underground facilitates various dreams and aspirations of ascendency. And yet, as this book has traced, ambivalence inevitably follows the initial optimism that planners and developers feel and that they help generate in the people who buy into the promise of the underground. Despite concerted efforts to package the underground in aesthetically acceptable ways, the city's subterranean spaces are rarely able to fully overcome the underground's physical qualities: darkness, dampness, and dirtiness. Even when successfully staged for comfort and style, the branded and expanded underground retains associations with poverty and immiseration that are not so easily papered over. And so the Metro is ridden begrudgingly and never as frequently as the municipality wishes; public parking garages go underutilized and subsequently unbuilt; and those living and working in basement conversions avoid hosting friends and find it more productive to take their meetings out of the office—all while dreaming of their next move, not just inward toward the center of it all but also upward. As Gabi made sure to note toward the end of our conversation, the kind of incorporation that the underground offered was, in the end, degraded. The sense of belonging it made possible was at best an approximation. "Sure, I was excited when I first bought my place, but I don't want to be living like this forever," Gabi began. "I have to use the Metro to be in the city. It's dirty. It's crowded and uncomfortable. It's hot in the summer, and you show up places sweaty," Gabi groused before making a basic observation that undercut the logic of how his building was packaged and sold to him: "Taking the Metro isn't the same as walking to the city center." The subtle indignities of the gentrified underground invariably tinge the kind of belonging the underground makes possible.

What this ethnography of the dreams and degradations of the urban underground makes clear is the need to account for the city's emergent verticality when conceptualizing what it means to belong to the city. At present, belonging has been theorized in strikingly horizontal terms. From Jürgen Habermas's notion of "the public" as a "parity of the educated" to Benedict Anderson's anchoring of "the nation" in a "deep horizontal comradeship," and from Victor Turner's notion of *communitas* as denoting "a communion of equal individuals . . . without which there could be no society" to Émile Durkheim's sense of the "collective effervescence" that binds a society together through shared affects, to belong has conventionally been understood to mean to share in a presence in which everyone is here together within the same horizontally delimited space.[3] And to be sure, the underground's renovated basements and cellars, transit stations and tunnels, kiosks and shops, and grave plots facilitate copresence in spades. These spaces enable more people to be "here," in the city center. And yet "here" is experienced with more nuance and variation than planners, developers, and users alike anticipated.[4]

Even as the gentrified and expanded underground has enabled more people to be "here"—that is, in the city center—it has not evoked a sense of horizontal comradeship or collective effervescence for its users; neither have those who utilize the underground experienced a sense of *communitas* or of participation in a shared public. This is because the "here" of the city is neither inhabited nor experienced upon the horizontal plane alone. By design, people's positionality relative to the city is increasingly as vertical—above and below the city's surface—as it is horizontal. The question of what it means to belong goes beyond whether one finds oneself "here" in the city or outside of it, within the gated community or on the outside looking in.[5] As those who have turned optimistically toward the underground eventually acknowledge, to be "here" in the center yet at the city's bottom is in fact to fall beneath those on top. Rather than standing shoulder to shoulder, equally "here" on a point along some shared horizontal plane, those at the bottom of the city experience a subsidiary sense of belonging vis-à-vis those on top, one marked by all of the affective indignities that being below implies. The dismissive attitude of those on top affected toward those on bottom not only aggravates but affirms these anxieties. It has been this book's central contention that the dreams and degradations of the gentrified urban underground have played a foundational role in shaping what it means to be middle class in twenty-first-century Bucharest.

The aesthetic repurposing of the underground that this book has traced remains emergent, its potential not yet fully explored. In the name of sustainability, planners and developers, architects and designers, as well as city residents have only begun to experiment with how the underground might be reimagined toward different ends. An estimated one million people are already believed to live beneath the streets of Beijing.[6] From Montreal to Seoul, subterranean pedestrian streets where people can shop, eat, and drink, as well as visit museums have become so extensive that the underground is officially promoted as a tourism destination in its own right.[7] Additionally, near Sydney's central business district, the elegantly executed Green Square Library and Plaza has extended public life downward, beneath the city.[8] While these spaces vary in quality and character, their scope continues to expand, and the number of people relying upon them grows daily. As cities develop further downward, and as the aesthetic repurposing of the underground matures, scholars and practitioners of the twenty-first-century city will need to become ever more attuned to the cultural politics of the city's verticality, asking not just who is here in the city and who has been pressed outside of it, but also who is on top and who is on bottom.

A Note on Method

The principal research for this book took place between 2014 and 2019. During that time, I spent a cumulative total of twelve months in Bucharest studying the gentrification and expansion of the city's literal underground. This fieldwork unfolded amidst my own transition from an unencumbered graduate student to the more demanding roles of professor and parent, which gave the research process for this book a different cadence relative to my previous one. The hollowed, ethnographic experience of having a long duration of uninterrupted fieldwork gave way to shorter trips that took place between teaching and parenting obligations. These trips occurred from June through July 2014, July through August 2015, January through April 2016, and May through June as well as December of 2017, 2018, and 2019.

Between trips, social media helped me to maintain key relationships while newspapers kept me abreast of major developments. Over the course of this time, I spoke only briefly with the unhoused persons and aid workers whom I had at first expected to be at the center of this book (see Preface). To bring the aesthetically repurposed underground into focus, I instead found myself talking at length with city planners, private developers, architects, real estate agents, renters, homeowners, landlords, revelers and ravers, academics, engineers, geologists, physicists, firefighters, police officers, Orthodox priests, municipal and ministry officials, boxing coaches, business executives and small business owners, advertising creatives, entrepreneurs, and interior designers. Each of these people was invested, in one way or another, in extending the lives of the middle classes downward, beneath the city's sidewalk. In short, to bring this transforming underground into view, I spoke with a wide range of actors. Well-trained and undeterred research assistants from the University of Bucharest helped me to arrange short, formal interviews that generally segued into invitations for longer follow-up meetings, site visits, and opportunities to shadow and observe. Cold calling gave way to snowballing as people offered to put me in touch with their colleagues, friends, and neighbors. The fieldwork occurred in a mix of Romanian and English. Many of the highly educated professionals and academics I interviewed were fluent in English, had done their degrees in English, and preferred using English with me. Most of my conversations with small business owners and entrepreneurs as well as visits to construction sites occurred in Romanian.

I approached the ethnographic writing as both a literary and practical challenge. Rendering twelve months of fieldwork into a monograph entails sensitive choices about representation that I have made only after careful consideration. I draw the quotations for this book from recorded and transcribed interviews and from detailed notes. While I relied on research assistants to do the transcribing, I am responsible for the translations. I translated and edited transcripts with great care and with reference to my fieldnotes in order to preserve the meaning of the ethnographic moment. To help give the book a clear narrative voice, I use ellipses within quotations to punctuate the moments when voices trail off pensively rather than to denote the omission of small phrases, repetitive information, or extraneous details. For the sake of analytical clarity, the material is arranged thematically rather than in the chronology of its collection. Each chapter focuses on a core set of interlocuters and cases. Temporal markers provide the reader with a sense of when the bulk of a given chapter's research occurred, however the analysis benefits from the totality of the fieldwork. To develop analytical nuance and clarity, on occasion I have moved observations and voices around in time to bring the perspective of certain characters into dialogue. Throughout the book, I use pseudonyms when referring to the people I interviewed and observed to protect their anonymity. In certain ethnographic vignettes, I have also obscured or changed minor details that are immaterial to the analysis but that could be used to reveal a person's identity. A core set of geographical references organizes much of the book. To make the text more accessible to readers unfamiliar with Bucharest specifically and the Romanian language more generally, I translate Piața Unirii and Piața Universității as Union Square and University Square, respectively. I otherwise use the Romanian names for specific places.

Whenever possible, I foreground the voices and perspectives of my informants. One consequence of this decision is that I do not frequently appear in this book. When I do appear, it is generally in service of the analysis by pressing someone for nuance or clarity. One unintended consequence of this choice is that I appear in the book to be more knowledgeable in the moment than I actually was. The simple fact is that the wide range of actors interviewed for this book meant that I often found myself in conversation with professionals whose technical expertise and vocabulary I did not share. In these moments I confronted the common anthropological experience of unfamiliarity and struggled to find my bearings in real time. On more than one occasion, I asked questions of people with technical backgrounds that missed the mark. Over the course of a decade of fieldwork and writing for this book, however, the work of correcting my missteps, assumptions, and naivetes has helped to achieve this book's final coherence.

Each chapter contains images intended to illustrate and extend the analysis. Except when indicated, photographs were taken by the author. Underground spaces, I quickly learned, pose practical challenges. Most relevant, longer camera exposures are required to compensate for low-light settings, which cause bodies in motion to blur. To minimize these practical complications, I took the majority of my images underground at off-peak hours. I am also sensitive to the politics of privacy in public spaces. To the extent possible, when selecting images of spaces in use, I selected photographs taken at a distance so that the identifiable features of people are at least partially obscured.

The long duration of this fieldwork allowed me to see both gentrification and development as ongoing processes rather than as singular events. Many of my field sites underwent substantive changes between the research and the completion of the writing for this book. For the most part, these changes were alluded to as potentialities during the fieldwork and appear as such in the book's narrative. For example, Chapter 1 draws upon interviews and site visits at Union Station's McDonald's restaurant that occurred in 2016. The restaurant then abruptly closed in 2019 over a leasing dispute. Chapter 2 draws upon interviews with Metro Kiosk owners that took place between 2015 and 2018. After thirty years of hosting kiosks inside of its Metro stations, the Metrorex Corporation canceled its contract with Sindomet Servcom SRL, the company that oversaw all kiosk leasing agreements. Amidst a cloud of scandal, Metrorex en mass cleared all kiosks from its stations in 2021. The bulk of the research on Old Town's basement-level nightclubs that appears in Chapter 4 occurred in the summer of 2015, before the tragic Club Colectiv fire shifted the model of these businesses while follow-up fieldwork in 2016 captured the aftermath (see Chapter 7). And as I write this very sentence,

Russia's ongoing invasion of Ukraine, which began in the winter of 2022, extends to Bucharest's civil protection shelters a renewed practicality that did not exist during my interviews and site visits, which occurred between 2017 and 2019 (see Chapter 9). The research for the book concluded in December of 2019, a few short months before the Covid-19 pandemic upended almost every aspect of everyday life. Taken together, the simple fact is that the gentrification of the underground is an unfinished and ongoing project. While these changes shift the underground's exact configuration and contours, they do not alter the underlying dialectic of aspirations and anxieties, dreams and degradations that animate the city's downward expansion. Books, in the end, are not newspapers. They do not offer up-to-the-moment details about current events. What this book does offer is deeper insights into how and to what effect a set of market and municipal actors have aesthetically repurposed and expanded their city's underground in the name of securing their city's future prosperity. While the pages you have just read concern Bucharest, the analysis broadly resonates with a number of other similarly positioned cities across Europe, Asia, and the Americas that are struggling with the challenges of over-development, even as they try to attract even more foreign investment.

NOTES

Introduction

1. Hénard, "Cities of the Future."

2. Webster called for the careful coordination of street and subsurface planning. He argued that this coordination required "municipal control by some official body with absolute power to finally approve the location of every sub-surface structure and to see that each is placed in the space assigned to it." Webster, "Subterranean Street Planning," 201.

3. Bell, "Air Rights," 250.

4. Eliade, *Sacred and the Profane*, 36.

5. Bell, "Air Rights," 250.

6. The proliferation of rail tunnels in cities such as Chicago and New York raised speculation about how the valuable land above those tunnels could be most effectively used. Bell continued: "One who looks at the map of downtown Chicago will be impressed by the immense amount of space occupied by railroad terminals and approaches. The iron band of the elevated railroad loop has been definitely broken through on the north at least, but there is another and broader band farther out where the railroads are entrenched apparently for all time. Speaking generally, the railroads use only the surface and one or two levels above and below the surface. The space above the railroad tracks not actually used for terminals is a complete economic waste. It is a sort of vacuum into which business is bound to press if it is given a chance." Bell, "Air Rights," 252.

7. "Park Avenue in New York City has already been mentioned, and no doubt there are similar stretches of tracks in every important city which lie near the center of the city and over which the air rights could be used. There is no doubt that it will not be many years before we will see these valuable air rights improved with beautiful buildings and perhaps boulevards. The method of appropriating the air rights over rail-road tracks depends entirely upon the character of the ownership of the property." Becker, "Subdividing the Air," 41. See also Goldschmidt, "Air Rights"; Comaroff, "On the Materialities of Air"; and Jose, "Hawa Khaana in Vasai Virar."

8. As Franck Billé notes, "The assumption that sovereignty extended upward and downward ad infinitum was only challenged when the technological means to colonize these upper and lower reaches became available." Billé, *Voluminous States*, 9–10. While extraction economies raise complicated questions about the lower reaches of sovereignty, the question of property rights in cities remains largely unaddressed; Billé, 9.

9. See Reynolds, *Underground Urbanism*, 48; also Delmastro, Lavagno, and Schranz, "Underground Urbanism," 104. As explored in Chapter 7 of this book, this opacity is in part due to the tremendous depth required to produce stable foundations for skyscrapers.

10. Wainwright, "Billionaires' Basements," in Graham, *Vertical*, 314.

11. Reynolds, *Underground Urbanism*, 48–50.

12. Simmel wrote: "For the eye [the bridge] stands in a much closer and much less fortuitous relationship to the banks that it connects than does, say, a house to its earth foundation, which disappears from sight beneath it. . . . Thus the door becomes the image of the boundary point at which human beings actually always stand or can stand . . . in contrast to the bridge which connects the finite with the finite." The earth foundation, and the cellar embedded within it, drop out of analytical view. Simmel, "Bridge and Door," 7–8. Important exceptions include Augé, *In the Metro*; Lemon, "MetroDogs"; Toth, *Mole People*; Fisch, *Anthropology of the Machine*; and Sadana, *Moving City*.

13. On the informal economy, see Bourgois, *In Search of Respect*; Venkatesh, *Off the Books*; and Contreras, *Stickup Kids*. On subversive politics, see Juris, *Networking Futures*; Graeber, *Direct Action*; Abufarha, *Making of a Human Bomb*; also Biehl, *Vita*.

14. For vulnerability, see Myradl, *Challenge to Affluence*; Wilson, *Truly Disadvantaged*; and Wacquant, "Urban Outcasts." For resourcefulness, see MacGaffey and Bazenguissa-Ganga, *Congo-Paris*; Simone, *For the City Yet to Come*; and Murray, *City of Extremes*.

15. See Simmel, "Bridge and Door"; Caldeira, *City of Walls*; Low, *Behind the Gates*; Rodgers, "'Disembedding' the City."

16. A horizontal perspective dominated by the coordinates of center and periphery has prevailed in discussions of inclusion and exclusion in the city since the Chicago school. See Park and Burgess, *City*; Williams, *Country and the City*; Castells, *Rise of the Network Society*; Sassen, *Global City*; and Appadurai, *Modernity at Large*.

17. Smith, *New Urban Frontier*; Zukin, "Gentrification"; Marcuse, "Gentrification, Abandonment, and Displacement"; and Harvey, "Flexible Accumulation through Urbanization."

18. This approach to the urban underground follows Henri Lefebvre's conceptualization of space as physical, mental, and social. As Lefebvre argued, "The aim is to discover or construct a theoretical unity between 'fields' which are apprehended separately, just as molecular, electromagnetic and gravitational forces are in physics. The fields we are concerned with are, first, the *physical*—nature, the Cosmos; secondly, the *mental*, including logical and formal abstractions; and, thirdly, the *social*. In other words, we are concerned with logico-epistemological space, the space of social practice." Lefebvre, *Production of Space*, 11–12.

19. Anthropologists use the language of "the field" and "field site" to demarcate the location of their research. The borders and boundaries demarcating fields and sites are, as others have shown, not straightforward spatial facts but instead cultural constructions. Gupta and Ferguson, "Discipline and Practice"; Bourdieu, "Site Effects."

20. Simmel, *Metropolis and Mental Life*; Pike, *Subterranean Cities*.

21. Pointing to deeply rooted Christian cosmologies, the Romanian philosopher Mircea Eliade noted that downward positionality is readily associated with the inferior and the profane and exists in opposition to the good, which is associated upward toward the heavens. Eliade, *Sacred and the Profane*. Applying as much to the bowels of the earth as to the depths of the body, Mikhail Bakhtin elaborated, the "downward turn" signals a move toward the grotesque and the obscene, the consequences of which can be socially destabilizing. Bakhtin, *Rabelais and His World*, 370. See also Eliade, *Old Man and the Bureaucrats*; Douglas, *Purity and Danger*.

22. Gandy, "The Paris Sewers and the Rationalization of Urban Space"; Céline, *Journey to the End of the Night*; Cărtărescu, *Nostalgia: A Novel*; Pike, *Metropolis on the Styx*.

23. As Rosalind Williams notes, advances in modern engineering since the nineteenth century have recast the underground from a site of hell conjured from mythology, to the steam, fire, grime, and noise of the first industrial revolution, and then to futurity and the technological sublime. Williams, *Notes on the Underground*, 97. See also Masco, "Life Underground."

24. For the characterization of basement housing as abject, see Engels, *The Housing Question*. 99. For the commodification of basement-level apartments for middle-class professionals, see Chapter 8 of this book.

25. Hewitt and Graham, "Vertical Cities."

26. Scholars in a variety of disciplines have similarly turned attention to the inequalities of inclusion, from economics (Piketty, *Capital in the Twenty-First Century*) to sociology (Burawoy, "Facing an Unequal World") and anthropology (Ong, "Graduated Sovereignty in South-East Asia"). While also bringing attention to matters of wealth accumulation and varied forms of citizenship, the aim here is to examine how unequal inclusion gets built into the city in ways that shape everyday life.

27. Important questions have been raised by spatial theorists about the three-dimensionality of territory: see Elden, "Secure the Volume"; Billé, *Voluminous States*; Graham, *Vertical*. The discussion has focused primarily upon questions of sovereignty. To the extent that issues of class are explored, the discussion has centered on the relationship between the street and the upper reaches of the city's penthouses. See Graham, "Luxified Skies"; O'Neill, Kevin L., and Fogarty-Valenzuela, "Verticality"; Harris, "Vertical Urbanisms"; Leitner et al., "Space Grabs."

28. O'Neill, "Up, Down, and Away," 140.

29. As James Ferguson notes, the word *place* has a double meaning, referring as much to a location in space as to a rank in a social order. Ferguson, *Global Shadows*, 6.

30. Grama, *Socialist Heritage*, 105.

31. Danta, "Ceaușescu's Bucharest."

32. Turnock, "Housing Policy in Romania"; Turnock, "The Planning of Rural Settlement in Romania."

33. Patron et al., "Conceptul strategic București 2035," 121.

34. On the location of privilege in Bucharest's center-north neighborhoods, see Chelcea, "Marginal Groups in Central Places." On the importance of density and public transit in socialist cities, see Crowley and Reid, *Socialist Spaces*; Deletant, *Ceaușescu and the Securitate*, 307.

35. O'Neill, "The Political Agency of Cityscapes."

36. Demekas and Khan, "The Romanian Economic Reform Program."

37. Lavinia, "Romanian Privatization," 430.

38. Harris, "Structural Adjustment and Romania," 2861, 2863.

39. On the percentage of people living below the poverty line, see Petrescu, "Romania Country Brief." For mortality rates, see Stuckler, King, and McKee, "Mass Privatisation and the Post-Communist Mortality Crisis."

40. Semo, "Une ville de chien"; "Romania's Capital Mayor"; McNeil, "Bucharest Journal."

41. Meyer, "Moment of Truth."

42. Dospinescu and Russo, "Romania," 17.

43. Alexe et al., "Social Impact of Emigration and Rural-Urban Migration in Central and Eastern Europe," 2; European Commission, "2021 Retrospective on Migration in Romania."

44. Dospinescu and Russo, "Romania," 14.

45. Perrin, "Romania as the Destination," 17–18.

46. The 2014 A. T. Kearney Index ranked Romania fifth among the most attractive outsourcing destinations in Europe. The 2012 *Times* Outsourcing Business Supplement ranked Romania sixth globally in the top ten emerging outsourcing destinations. A 2014 study by KPMG also ranked Romania as the fastest growing market in the European Union in terms of IT outsourcing services. Vigroux, "Preferred Outsourcing Destination." Regarding the amount of FDI received by Romania, see Estrin and Uvalic, "FDI into Transition Economies," 289. Roughly half of the FDI pouring into Romania was directed toward Bucharest; Danciu and Gruiescu, "Evolution of FDI," 101.

47. Teodorescu and Tudor, "Demographic Analysis of the Bucharest-Ilfov Region," 595

48. Dumitrache, Zamfir, and Nae, "The Urban Nexus," 47; Alexe et al., "Social Impact of Emigration and Rural-Urban Migration in Central and Eastern Europe," 2.

49. The number of new buildings under construction in Bucharest grew to one thousand in 2002, only to nearly double to 1,800 by 2006; see Chelcea, Popescu, and Cristea, "Who Are the Gentrifiers," 120. Following the global financial crisis, Bucharest's modern office stock nearly doubled again from 1.52 million square meters in 2012 to 2.85 million square meters by 2018. See Colliers International, "Romania"; Alexe, "CBRE," also Ion, "Public Funding," 173.

50. Toma, "Room without a View." In response, the Ministry of Health ordered that all rooms in a housing unit receive sunlight for at least two hours every day, prompting a dean at the University of Architecture (UAUIM) to quip that, under such a stipulation, 90 percent of the buildings in Bucharest were uninhabitable.

51. Pen and Hoogerbrugge, "Economic Vitality of Bucharest," 14.

52. Ellyatt, "'Tiger' of Eastern Europe"; Timu and Vilcu, "Romanian GDP Growth Quickens."

53. For "fuzzy," see Wacquant, "Making Class," 51. For "breathtakingly vague," see Ost, "Stuck in the Past," 614.

54. Yeh, *Passing*; Warner, "Publics and Counterpublics." For an extended discussion, see Chapter 8 of this text.

55. For a discussion on how Bucharest's ascendant middle classes, rather than its downwardly mobile workers and poor, emerged as the principled subject of state and private investment alike, see Chelcea and Druță, "Zombie Socialism," 530.

56. As Franz Boas noted with reference to Epicurus, aesthetics entail a desire to order and arrange phenomenon into a clear and desirable system. Boas, "The Study of Geography," 139.

57. Eurostat, "Earnings Statistics"; Romania-Insider, "Average Net Income."

58. Eurostat, "Earnings Statistics."

59. Dreger et al., "Wage and Income Inequality," 27. This income disparity was also reflected in minimum-wage workers. The minimum wage in Luxembourg, for example, was €1,923 per month, while it was only €218 per month in Romania—the lowest in the EU next to Macedonia (€214) and Albania (€157). Eurostat, "Minimum Wage Statistics."

60. Chelcea, "Post-Socialist Acceleration," 349.

61. Colliers International, "Research & Forecast Report," 28. Rising standards of living and the increased availability of credit for automobiles and housing drove strong urban growth on the periphery of Bucharest. New residential neighborhoods developed along the outskirts of the city, especially in the wealthy northern part of Bucharest. Pen and Hoogerbrugge, "Economic Vitality of Bucharest," 14. These affordably priced, newly constructed developments came readymade with modern design finishes unlike the socialist-era housing blocs of inner residential neighborhoods, which required costly renovations to be brought up to so-called

"European standards." For a discussion on the shifting material culture of housing in post-socialist Eastern Europe, see Fehérváry, *Politics in Color and Concrete* and Nadkarni, *Remains of Socialism.* This line of analysis is picked up and applied to Bucharest specifically in Negru, *De la stradă.*

62. Synergetics Corporation et al., "Plan integrat," 39.

63. TomTom International BV, "TomTom Traffic Index."

64. Between 2005 and 2011, the Bucharest Ilfov region reported average annual concentrations of nitrogen dioxide in the surrounding air in excess of the annual limit value for human health ($40\mu g/m^3$). Compared to other European capitals, Bucharest recorded the highest concentrations of nitrogen oxides every year, exceeding values recorded at monitoring stations in European capitals such as Vilnius, Stockholm, Paris, and Sofia, marking Bucharest among the most polluted European capitals. ADRBI, "Planul de dezvoltare," 149–50. See also "Most Polluted Cities."

65. "Mii de mașini vechi abandonate." Romania's socialist-era planners always assumed the predominance of public transportation and did not accommodate mass personal automobility into their development plans. Blocks built before the 1990s have only a handful of parking places. Moga, "Mall-urile au creat." For further discussion, see Chapter 5 of this book.

66. Rom Engineering Ltd and AVENSA Consulting SRL, "Planul de mobilitate," 29.

67. In the Master Plan for Transport, planners predicted "that the number of inhabitants will increase moderately in the metropolis of Bucharest by 1.1% in the period 2005–2015 and by 1.9% in the period 2005–2027. It is estimated that real GDP will continue to develop at a rate that will be above the national trend, resulting in average growth rates of 6.1% (2005–2015) and 5.4% respectively (2005–2027). It is also forecast that the number of employees will increase above the national average, with a change in percentage of 4.8% by 2015 and of 8.7% by 2027. In the context of the projected favorable economic conditions, the motorization is expected to develop dynamically: 461 cars per 1000 inhabitants in 2015 and 579 cars per 1000 inhabitants in 2027." Consiliul General al Municipiului București, "Master planului," 42.

68. Bucharest's "dysfunction will accentuate exponentially, and will contribute to the declining prestige of Bucharest, which wants to be a European capital, and stagnate and regress the economy, which will grow a comparative gap with other European states," municipal planners noted. Patron et al., "Conceptul strategic București 2035," 173. The Strategic Concept for Bucharest 2035 continued, "Non-intervention and unresolved acute dysfunctionalities of the landscape, made coherent and through a concerted strategic intervention, will have repercussions not only on the environment and urban and human health, but also on the economic and social field, with possible financial sanctions for non-compliance with European requirements." Patron et al., 173, also 181.

69. Ignat, "Paianjenul Metrorex."

70. The initial press release estimated the cost of the project to be €7 billion. Serviciul pentru Relația cu Mass-Media, "Conferinta." This figure was later revised to €8.5 billion. Ignat, "Paianjenul Metrorex."

71. See Belzberg, *Children Underground.* The association between the Metro and poverty is true in other post-socialist cities. Sherouse, "Where the Sidewalk Ends"; Lemon, "Talking Transit."

72. The unavoidable presence of homelessness in, below, and around the Gara de Nord Station has its origins in the pronatalist policies of socialist-era Romania. In the 1980s, the dictator Nicolae Ceaușescu outlawed birth control in the name of growing Romania's

workforce and ultimately building socialism. See Kligman, *Politics of Duplicity*; Nelson, *Romania's Abandoned Children*. As these policies took effect against the backdrop of tightening austerity, birthrates quickly outpaced the capacity of parents to care for their children. Households already living hand-to-mouth turned over responsibility for approximately 150,000 children they could not afford and did not necessarily want to poorly funded state-run orphanages. Kligman, *Politics of Duplicity*; Morrison, "Ceausescu's Legacy"; Gloviczki, "Ceausescu's Children," 117. After the end of Romanian socialism, American television viewers glimpsed the deplorable conditions of these orphanages by way of the ABC News segment, "Shame of a Nation." Commonly referred to as "street children" (*copii străzii*), homeless youths had fled deplorable state-run orphanages during Romania's economic collapse in the 1990s to take their chances in Bucharest, often seeking shelter in the Metro.

73. See Lancione, "Weird Exoskeletons"; Lancione, "Underground Inscriptions."

74. PMB, "Strategia de parcare," 49.

75. Racu, "Cât costă să faci."

76. "Moartea este un lux"; Ivanov, "Piaţa locurilor"; Toea, "Un loc de veci."

77. Paris provided the elites of nineteenth-century Bucharest with a recognizable model for orienting Romania westward, toward Europe. "The construction of Romanian identity involved a constant search for prestigious and legitimizing cultural models elsewhere, the most prominent and durable of which was not just the French paradigm in general, but particularly Paris as the cultural capital of nineteenth-century Europe." Spiridon, "European Self," 151. Bucharest's leading families actively sought to cultivate durable ties with Paris. They enrolled their youth in French universities, kept abreast of French political life, and were regularly seen in Parisian literary salons. Spiridon, "The Fate of a Stereotype," 272. As the Paris educated returned to Bucharest, they brought with them aesthetics and artistic sensibilities internalized while abroad. They hired French and Swiss architects whose designs reflected the Parisian Beaux-Arts school, gardeners for their landscaping, as well as carpenters who designed neo-Baroque bookcases upon which their French volumes could rest. Brzostek, "Romania's Peculiar Way," 111. Through the staged materiality of home architecture and interior design, Bucharest's elites imagined themselves as transforming their city into "Little Paris." The lack of formal planning, however, rendered Bucharest unreadable from a Western perspective. Unlike Paris, Bucharest did not develop out of a medieval city with a clearly defined central market but rather took shape at the confluence of several artisanal settlements (*târg*). Brzostek, 118. "Nineteenth-century French travelers visiting Bucharest questioned its status: 'Is Bucharest a village or a city?' 'Where is the city's center?' But Bucharest could not be understood in the same terms as Paris, since historically the Romanian city was formed as a gathering of several villages, while Paris's urban existence, like that of other Western capitals, started at the center of the medieval city and then expanded concentrically. The absence of a center translates into a long list of observations about disorder—lack of systematic arrangement, messiness, and disruption—that betrays the travelers' inability to shift mentality from Paris to Bucharest." Verona, "Bucharest at the Crossroads," 276.

78. "Bucharest is an Eastern European capital that has many gaps compared to the large developed European capitals. . . . It is in direct competition with Belgrade, Budapest, Sofia, or Istanbul . . . even if it cannot compete directly (and not because of the lack of proximity, but taking into account the large gaps in the level of development) with Stockholm or Vienna or Paris, Bucharest must relate to them, especially to what they have undertaken to achieve

development and in the way in which they capitalize or build new distinctive competences." Patron et al., "Conceptul strategic Bucureşti 2035," 294.

79. Synergetics Corporation et al., "Plan integrat," 140.

80. Quoting Michel Foucault, Gilles Deleuze defines the diagram as a "'functioning, abstracted from any obstacle . . . or fiction [and which] must be detached from any specific use.' The *diagram* is no longer an auditory or visual archive but a map, a cartography that is coextensive with the whole social field. It is an abstract machine." Deleuze, *Foucault*, 34. For Deleuze, "the diagram acts as a non-unifying immanent cause that is coextensive with the whole social field: the abstract machine is like the cause of the concrete assemblages that execute its relations; and these relations between forces take place 'not above' but within the very tissue of the assemblages they produce." Deleuze, 37. Taking the idea of diagram and linking it to the governance of cities, Osborne and Rose write: "Of course, one could be literal about the diagram: there are so many plans, schemes, drawings, stories, myths, and programmes. . . . But the urban diagrams that concern us are to be discerned within all these actual schema, immanent in them, perhaps providing their historical a priori: the various abstract cities which, at different moments, distribute various attempts to understand and intervene into concrete urban space, time, and existence. These diagrams are neither models nor Weberian ideal-types but operative rationales. Each diagram depicts and projects a certain 'truth' of the city which underpins an array of attempts to make urban existence both more like and less like a city." Osborne and Rose, "Governing Cities," 738. On Bucharest's Parisian aspirations, see Synergetics Corporation et al., "Plan integrat," 140. In writing about the myth of Bucharest as the "Little Paris," the Romanian historian Lucian Boia notes: "Despite a number of Parisian-style buildings from the last decades of the nineteenth century, Bucharest as a whole does not resemble Paris. Something of the Parisian lifestyle characterized the behavior of an elite, and certain corners of the Bucharest cityscape acquired a Parisian atmosphere. However, the greater part of the population lives far from the French model." Boia, *History and Myth*, 162.

81. Bruce Grant borrows the term "edifice complex" from the architectural critic Deyan Sudjic to describe a government's strategic efforts to use ambitious building projects to exorcise the past and to pursue recognition on the world stage. Grant, "Edifice Complex," 505. See also Sudjic, *Edifice Complex*. Using the case of Baku, Azerbaijan, Grant charts the mobilization of a soaring verticality, articulated through glass and steel towers, to "conjure an image of the 'technological sublime'—a world where the skyline carries the citizen's imagination above and ultimately away from the known life of the sidewalk." Grant, 505. Grant's work aligns with Aihwa Ong's notion of hyperbuilding, which Ong defines as an intense process of building to project an urban profile and to signal a city's political and economic aspirations. Ong, "Hyperbuilding," 206–7. To be sure, municipal and market actors in Bucharest seek world recognition in part through a rising skyline shaped by office and residential towers (see O'Neill, "Up, Down, and Away"), but as noted previously, planning documents also anchor claims to Bucharest's economic competitiveness on a world stage by styling its city center as a "Little Paris."

82. Vintilă, *Bucureştiul subpământean*, 11.

83. Vintilă, 76. After the Princely Courts were abandoned by the wealthy, the cellars of the ruined palaces were taken over by society's most marginal figures—the poor, the drunk, and the ethnically Roma—who were derisively referred to as "The Beaux of the Old Court." The Old Court became a site of drunken debauchery, as captured by the novelist Mateiu

Caragiale, who wrote in *Craii de curtea-veche*: "Like the rest of the town, the Court had more than once risen from ashes, and it must have covered quite a stretch, judging by the vaulted ruins to be found beneath much of the neighborhood—even under this our bistro, in fact. The overall design of the Court was not hard to imagine, as it largely followed the monasterial pattern, with building clusters to accommodate a host of commoners and gypsies." Caragiale, *Gallants of the Old Court*, 23.

84. Closely connected to the inns of Bucharest are its pubs, equipped with deep, cool wine cellars, which supported large feasts. The large number of vineyards and orchards covering the hills of Bucharest made such cellars a common feature of the city's architecture. Vintilă, *Bucureștiul subpământean*, 34. For a discussion of the architectural importance of the vaulted brick cellar for country manors, see Burcuș, "Vizită."

85. Gheorghiu, "Arhitectura subterană." City residents often speak of discovering such tunnels during childhood play or following major earthquakes. As one written account noted, the author "followed an underground road to the dim meadow where my ancestors fled in front of the invaders. . . . I found the end of this underground road under the surrounding wall following a landslide caused by the 1977 earthquake." Predescu, *Vremuri vechi Bucureștene*, 103.

86. Săveanu, *Enigmele Bucureștilor*, 18–23. This was captured also by Mateiu Caragiale who wrote of a wealthy merchant, "As you will have seen, the closets in my study are walled in and hidden behind heavy drapery, so as one shouldn't easily notice they have no bottom and open down into [an escape] tunnel." Caragiale, *Gallants of the Old Court*, 115.

87. Arhivele Naționale României, "Țara Românească Dosar No. 169."

88. Dragomir, *Un București mai puțin cunoscut*, 83; Filip, *Bellu*.

89. Iancu, "Utopia Bucureștilor," 15–16. See also Machedon and Scoffham, *Romanian Modernism*.

90. Banister, "Transport in Romania," 261.

91. Beznilă, "Stația piața Unirii," 19.

92. Popescu and Feraru, "Convorbire," 12; see also Novitchi, "Metroul din București," 11; Hatherley, *Landscapes of Communism*, 250–309.

93. According to accounts, the history of this plan began immediately after the Soviet invasion of Czechoslovakia in August 1968. Following the invasion, Ceaușescu ordered the drafting of a plan to evacuate him from Bucharest during a moment of instability. Curea, "Planul Z"; Răduță, "Adevărul despre tunelurile subterane."

94. One journalistic account of Bucharest's socialist era escape tunnels reads as follows: "We have an account from an army major, special forces. He explored the tunnels without a plan or idea of what they'd found; entered through an armored door; in the first tunnel was a well-organized basement with a barracks probably for Ceaușescu's guard; notes beds and radiators; every room has its own color; another is a rations room with food, tools, and bottles of pure oxygen; flashlights with imported alkaline batteries; around 300 pistols and bullets; found a ventilation opening with 50 grenades; bunker/office with 'improved conditions.' The cabinet has porcelain dishes, crystal glass, bottles of whisky and a refrigerator with food dated 12/21." Alexandrescu, "Orașul."

95. Michel de Certeau defines strategy as "the calculation (or manipulation) of power relationships that becomes possible as soon as a subject with will and power (a business, an army, a city, a scientific institution) can be isolated. It postulates a place that can be delimited as its own and serve as the base from which relations with an exteriority composed of targets and threats (customers or competitors, enemies, the country surrounding the city, objectives

and objects of research, etc.) can be managed. As in management, every 'strategic' rationalization seeks first of all to distinguish its 'own' place, that is, the place of its own power and will, from an 'environment.' A cartesian attitude, if you wish: it is an effort to delimit one's own place in a world bewitched by the invisible powers of the Other." Certeau, *Practice of Everyday Life*, 36. This book is interested first and foremost in the underground as a strategic site, one that is produced and staged for an imagined subject and one that seeks to shape that subject in desired ways.

96. ACUUS, "Cooperation Agreement"; Admiraal and Suri, *Think Deep*; ITA-AITES, "Urban Underground Space."

97. Killada and Raju, "World's Top Economies"; Mao et al., "Global Urban Subway Development."

98. McNeill, "Volumetric Urbanism"; Vähäaho, "Underground Space Planning in Helsinki."

99. WSP Global Inc., "Taking Urban Development Underground."

100. Frampton, Wong, and Solomon, *Cities Without Ground*.

101. Adler, "The Next Frontier."

Chapter 1

1. Unchecked development guided by the desires of foreign investors rather than planning regulations had led to unsustainable population and structural density within the city as well as uncoordinated suburban growth at its periphery. Taken together, city planners warned of a "structural and spatial fracture between the city and its metropolitan territory." Patron et al., "Conceptul strategic Bucureşti 2035," 88.

2. The expanded Metro projected outward as far as Bolintin to the west, Buftea and Mogoşoaia to the north, Berceni to the south, and Pantelimon to the east. Rotariu, "Cum va arăta." Metro systems have long moved significant populations on a daily basis. Lewis Mumford, writing about the New York City subway in the early twentieth century, described the role of metros and subways as being "mechanically to mobilize and disperse, night and morning, the inhabitants of the metropolis. . . . Though such transportation systems open up new areas on the outskirts of the city, they but thicken the crowding at the center." Mumford, *Culture of Cities*, 238–39.

3. Rom Engineering Ltd and AVENSA Consulting SRL, "Planul de mobilitate," 173.

4. See Woodhouse, "What Penn Station's $6 Billion Makeover Means for NYC"; Chrisafis, "Paris Hopes €1bn Revamp of Les Halles Can Become City's 'Beating Heart.'"; Topham, "Final Piece in £700m Overhaul of Bank Tube Station in London Opens to Public."

5. Enache, "Metrorex."

6. For the language of "misery and chaos," see Ivanov, "Metrorex." Others have documented the declining prestige of metros specifically and public transit more generally in postsocialist contexts. See Sherouse, "Where the Sidewalk Ends," 455; Lemon, "Talking Transit."

7. Böhme, "Art of the Stage Set," 2. Taking his cues from the theater, Böhme conceptualizes the atmosphere of a place from the perspective of stage design. He describes an atmosphere as a character or feeling communicated to an audience through the manipulation of the material conditions of sound and light. "Atmospheres are totalities," Böhme writes, "atmospheres imbue everything, they tinge the whole of the world or a view, they bathe everything in a certain light, unify a diversity of impressions in a single emotive state." Böhme, 2. Böhme uses the language of atmosphere and staged materiality to chart the development of what he calls

"aesthetic capitalism," in which commodities are produced for the sake of staging our lives rather than for the fulfillment of basic needs via their consumption. Böhme, *Critique of Aesthetic Capitalism*. I draw upon Böhme's work to advance the claim that class differentiation and inequality are produced and affected at the level of staged materiality.

8. Răduță, "Metrorex."

9. "The conception of atmospheres," Böhme notes, has its origin in reception aesthetics. Atmospheres are apprehended as powers that affect the subject, and they have the tendency to induce in the subject a characteristic mood. Böhme, "Art of the Stage Set," 2. Understood through the lens of reception, talk of atmosphere appears entirely subjective if not ungrounded. This is why, following Böhme, this chapter approaches the matter of atmosphere from the perspective of production aesthetics. "It is the art of the stage set which rids atmospheres of the odor of the irrational," Böhme argues. "For the stage set artist must relate them to a wider audience, which can experience the atmosphere generated on the stage in, by and large, the same way." Böhme, 3. This chapter explores how, by working upon the staged materiality of the underground, market and municipal actors produce a mood or feeling underground that resonates compellingly with the aspirations of middle-class audiences.

10. On June 16, 1995, McDonald's opened its first restaurant in Romania on Union Square. Voinea and Delcea, "Interviu: Povestea venirii McDonald's în România." McDonald's subsequently opened its Union Station restaurant with a twenty-year lease agreement. In an effort to further capitalize on the valuable retail space, Metrorex chose not to renew this lease, prompting the Union Station McDonald's to close in 2019. "McDonald's a închis restaurantul," 2019. Between 2000 and 2008, the franchise also operated a financially successful McDonald's inside of the University Station vestibule. Its lease was similarly allowed to expire amidst plans for a wholistic renovation. Although these particular restaurants closed, their staged materiality is emblematic of broadly used strategies to aestheticize the underground.

11. In *The McDonaldization of Society*, Ritzer argues that the standardization principles that enable McDonald's to reproduce the Big Mac at a global scale lead to cultural homogenization. Others have taken up this thesis to argue that McDonaldization produces sameness not only within its restaurants but also across the high streets and main squares in which global brands like McDonald's operate worldwide. See Featherstone, *Undoing Culture*; Tomlinson, *Globalization and Culture*.

12. Again, the modern underground is not abject. The underground can be aestheticized to a number of pleasing ends. As others have noted elsewhere, Metro stations have taken on a museum-like elegance, as in Moscow, and have evoked domestic comfort, as with the initial stations of the London Tube. For a photographic account of the Moscow Metro, see Kuznetsov et al., *Hidden Urbanism*. For an account of the design aesthetic of initial London Tube stations, see Pike, *Subterranean Cities*. Upon their opening, commentators in Bucharest praised its Metro stations for their modern design and "quality ambiance." Popescu and Feraru, "Convorbire," 12.

13. Management did acknowledge that the spatial specificity of the underground complicated the maintenance of the station's restaurant. The Bucharest Metro does not have service elevators to move fresh food underground or to bring trash back aboveground. Built in the socialist era, commercial infrastructure was never factored into the initial design of Bucharest's stations. McDonald's instead relied upon a dedicated train cart to deliver fresh supplies and to remove waste materials from the Metro by way of the platform. Employees manually

carried food and waste up and down the stairs connecting the station's platform and its vestibule.

14. Segmentation, in this instance, is "supple" rather than "rigid." Deleuze and Guattari, *A Thousand Plateaus*, 210–11. While McDonald's management tailored the experience of each restaurant for an identified customer segment, customers segmented themselves by choosing the restaurant experience that best resonated with their needs at a given moment. Customers were free to make different decisions at different times.

15. There is a temporal nuance to the Union Square McDonald's. It is located next to the Old Town nightlife district, and it is open twenty-four hours a day. In the middle of the night, long after the professional workday has ended, the Union Square McDonald's caters to those looking to sober up or to refuel before heading back out onto the dance floor. The college student or entry-level office worker who ordinarily stuck to the McDonald's belowground during the workday might, for example, have hung out at the aboveground restaurant on their nights and weekends. The Union Station restaurant, for its part, closed along with the Metro system at 11 p.m.

16. On the marketing of minimalism to the middle classes, Kyle Chayka writes that minimalism, "isn't necessarily a voluntary personal choice but [rather] an inevitable societal and cultural shift responding to the experience of living through the 2000s. Up through the twentieth century, material accumulation and stability made sense as forms of security . . . [but with] crisis following crisis; flexibility and mobility now feel safer than being static, another reason that owning less looks more and more attractive;" Chayka, *Longing for Less*, 10–11. The less-is-more minimalism installed underground to cater to the middle classes is explored in greater depth in Chapter 2 of this book.

17. The ability to manipulate temporal relations underground is a reoccurring theme among municipal and market actors alike. See Chapter 4 of this book on clubbing, and Chapter 8 on advertising.

18. Vintilă, *Bucureștiul subpământean*, 30. See also the Introduction of this book.

19. Vintilă, 32–34.

20. Elsewhere I have written about the rise of multimillionaires in Bucharest and their preference for living, working, and socializing above the city, in office and residential towers. O'Neill, "Up, Down, and Away."

21. In the early twentieth century, Romania's architecture journal, *Arhitectura*, attempted to identify and document a traditional Romanian architectural style. See Stănescu, "Case vechi Românești." These efforts drew attention to the spacious, vaulted cellars of Bucharest's grand villas, where windows were often adorned with artistically rendered iron gates. Trăjănescu, "Casa din strada Labirint." More than storage areas, the cellars of Romanian villas were lived spaces, where household duties such as food preparation and laundry were carried out below while the more social functions of receiving guests occurred on aboveground floors. Smărăndescu, "Casa Cotescu," 55; also Ciortan, "Casa Octav Gușerescu."

22. In his casting of the underground as a site of "other kinds of possibilities," Lorenzo conceptualized basements and cellars as what Michel Foucault called heterotopias, or spaces of juxtaposition. Foucault, "Of Other Spaces." The term "heterotopia" has its origins in medicine and refers to tissue that develops at a place other than its usual location. The tissue is not diseased or particularly dangerous but merely placed elsewhere, dislocated. Johnson, "Unravelling Foucault's 'Different Spaces,'" 77. Lorenzo spoke of Bucharest's basements and cellars as a kind of heterotopia, a dislocated space that effects a place and time of its own.

23. Basements and cellars have a rich history of romantic theorization. "Up near the roof all our thoughts are clear," writes Gaston Bachelard; "as for the cellar . . . it is first and foremost the dark entity of the house, the one that partakes of subterranean forces. When we dream there, we are in harmony with the irrationality of the depths." Bachelard, *Poetics of Space*, 18. Further on, Bachelard adds, "to go down to the cellar is to dream, it is losing oneself in the distant corridors of an obscure etymology, looking for treasures that cannot be found in words." Bachelard, 147. Mircea Eliade writes similarly of cellars as spaces of fantasy and escape. Eliade, *Old Man and the Bureaucrats*.

24. Reynolds, *Underground Urbanism*, 30. "Urban history is replete with extraordinary examples of the vast geologic work necessary to create it," Stephen Graham writes. "Mexico City, for example, which has sunk 10 meters in the last century, was built on the bed of a huge drained lake on the site of the Aztec capital of Tenochtitlan. Downtown Chicago, meanwhile, is the product of a nineteenth-century engineering project to raise the city's ground level above the swamps where it was located. In Manhattan, meanwhile, the vast grid-scape that was constructed across the whole island between the eighteenth and twentieth centuries involved more than just the surveying and laying out of the grid and its subsequent development. Right across the island, huge engineering works and earthworks were necessary to erase Manhattan's naturally hilly landscape by gnawing away at the uplands and using the material thus created to fill in the lowlands and valleys." Graham, *Vertical*, 285–86.

25. As Wilmott notes with reference to Hong Kong, "Measuring height is both situated and relative, since landscapes rise and sink, the tide washes up and down, and the oceans are not consistently level across the world. In Hong Kong, this meant that in 1866 a copper bolt was driven into the pier of the Naval Dockyard by the crew of the survey vessel, HMS *Rifelman*, to determine a consistent sea level for the measurement of the hills." Wilmott, "Surface," 149. The intervention here, in Bucharest, is to bring into view the commodification of the surface level, noting how individual property developers manipulate the surface for commercial ends.

26. As city planners explained to me, the developability of land parcels depends upon two constraints: the POT, which refers to the parameters of a proposed building's footprint, and the CUT, which refers to the gross built area allowed over the building's POT. For example, if a land plot has a POT of 5,000 m^2 and a CUT of 3, a contractor can build to a maximum 15,000 m^2. By law, the calculation of the CUT does not consider underground levels, allowing developers to increase their profits by developing downward.

27. As Dan noted, the development of basement apartment units pertained to central and semi-peripheral neighborhoods where available land was both expensive and scarce. In the city center, municipal requirements to develop parking spaces for each apartment unit already compelled developers to incorporate basement garages into their projects (see Chapter 5). Adding an additional basement level posed only a minor additional cost. Along the periphery, where land was both available and cheap, there was little incentive to develop downward instead of outward.

28. At the time of writing, the venue had turned from a café to a sushi restaurant. Though the color pallet had shifted, the overall design strategy remained the same.

29. In the writings of Charles Baudelaire, the café and restaurant are portrayed as popular haunts of the flaneur, providing an acceptable perch for non-ambulatory forms of flânerie. Smart, "Digesting the Modern Diet,"166. The café emerged as a privileged site for reading the transformations of experience within the modern metropolis. Frisby, "Analysing Modernity," 13. This makes sense given the role credited to cafés in producing modern public life. Haber-

mas, *Structural Transformation*, 30. Within this humanistic tradition of urban studies, the café terrace provides the opportunity to hone observational skills and to come to know a city exhaustively. Perec, *Attempt at Exhausting*. Once submerged belowground, even partially, the café's vantage onto street life as well as onto capitalist modernity becomes degraded.

Chapter 2

1. The company Sindomet Comserv SRL, which is managed by the Metro Union, received the contract to administer the Metro's advertising and commercial spaces in 1990. Stoica and Burlă, "Nababii din subteran."

2. Sindomet's contract expired in 2018 and, amidst great controversy, Metrorex intensified the aesthetic intervention into its stations in 2021 by clearing out commercial kiosks entirely. Cicovschi, "Metrorex începe evacuarea spațiilor comerciale de la metrou . . ."

3. At the time of this research, rental prices for commercial kiosks varied depending upon the station. In the most expensive stations such as Victoriei, Union, or the Gara de Nord monthly rental prices reached an estimated €40 per square meter. Stoica and Burlă, "Nababii din subteran."

4. Commenting on the quality and character of Metro kiosks, the then General Manager of the Metrorex corporation quipped glibly to reporters, "You think you're in Istanbul." "Adio covrigilor." For calls to introduce national and multinational brands to the Metro, see Ivanov, "Metrorex a rupt contractul"; Enache, "Metrorex vrea să rezilieze." Outside observers confirm as much, with one writing that Bucharest's Metro expressed "levels of squalor that Londoners or New Yorkers will find familiar, with surfaces mostly filthy." Hatherley, *Landscapes of Communism*, 295.

5. In developing his notion of "the right to the city," Henri Lefebvre criticized at length the commodification of urban life. "We already know the double character of the capitalist city: place of consumption and consumption of place. Businesses densify in the centre, and attract expensive shops, luxury foodstuffs and products. . . . It tends to absorb use value in exchange and exchange value." Lefebvre, *Writings on Cities*, 170. While exchange value became the predominant value guiding the production of space upon the city's surface, the demands of global capital accumulation did not achieve the same vise-like grip upon the production of space underground. Bucharest's Metro system, notably, operates under the control of the Ministry of Transportation. It is a state space oriented to public transportation and also to securitization (see Chapter 9). Whereas the city above conforms more closely to the needs of capital, companies wishing to do business underground must "fit" within the constraints of the Metro system.

6. In the turbulent decade after the end of socialism, Western reformers had pegged the successful transition of Eastern Europe toward a market-driven society on the cultivation of an entrepreneurial spirit. Ghodsee and Orenstein, *Taking Stock of Shock*, 49. Building upon the work of Nikolas Rose (Rose, "Governing the Enterprising Self"), ethnographers documented the reinvention of personhood required for workers to meet market demands. Workers had to become more flexible in their skillset, responsible for their careers, and able to manage new levels of risk. Dunn, *Privatizing Poland*, 22. See also Makovicky, *Neoliberalism, Personhood, and Postsocialism*; Yurchak, "Russian Neoliberal." This chapter explores one way in which the entrepreneurial ambitions of the middle classes were pressed underground and into the auxiliary spaces of Metro kiosks as well as basements and cellars as foreign investment flowed across central Bucharest, driving up the cost of real estate.

7. At the time of research, 5 to Go operated inside University, Grozăvești, Nicolae Grigorescu, and Gara de Nord Stations.

8. Workers have long internalized market pressures to function ever more efficiently by ingesting products that promise acceleration, from caffeine and high-energy drinks to Adderall and methamphetamines. See Sedgewick, *Coffeeland*; Pine, "Economy of Speed."

9. In her research on homelessness in New York City, Joanne Passaro notes the dramatic social distances between extreme wealth and poverty that must be negotiated while conducting fieldwork in a major metropolis. Passaro observes how subways collapse the distances between centers and peripheries, requiring shifts in dress and comportment as one moves through one's day and across the city. Passaro, "'You Can't Take the Subway . . .'," 152–54. As this chapter shows, subways also shape the verticality of the city, creating a subterranean tier whose aesthetic and class dynamics contrast productively, in this instance, with the city center above it.

10. "Adio covrigilor."

11. Every kiosk owner with whom I spoke declined to disclose their monthly rent, citing concerns that their lease would not be renewed if they violated their confidentiality clause.

12. For the €10 figure, see "Adio covrigilor"; for the €8 estimate, see Stoica and Burlă, "Nababii din subteran." Anthropologists have often contrasted their knowledge production against journalism. As Liisa Malkki argues, "setting up a binary contrast between anthropological and journalistic modes of knowledge production, even provisionally, unduly homogenizes and simplifies both kinds of practices." Malkki, "News and Culture," 94. Malkki encourages an anthropological engagement with journalism to enhance ethnographic knowledge production. Rather than redoubling the fact-finding efforts of journalists, this study has instead drawn upon them, which afforded time in the field to pursue new directions.

13. Ghica, "Mafia de la metrou."

14. "Adio Covrigilor."

15. "Time and space, the cyclical [organic] and the linear [social practices] exert a reciprocal action: they measure themselves against one another; each one makes itself and is made a measuring-measure; everything is cyclical repetition through linear repetitions," wrote Henri Lefebvre. *Rhythmanalysis*, 18. For Lefebvre, urban spaces achieve their distinctive character in part temporally, through the assemblage of their different "beats," or the cyclical and linear rhythms that reverberate and collide upon them. Crang, "Rhythms of the City," 189. The underground, this chapter emphasizes, is not just a configuration of a kind of space that contrasts with the city surface; it is the production of a kind of time whose distinctive rhythms impact the pace of social practices, from commuting to the entrepreneurial effort of running a kiosk.

16. 5 to Go, "Franchise."

17. Corruption is a longstanding and widespread problem in Romanian politics. And unsurprisingly, talk of corruption circulates through the underground. Between 2000 and 2016, the Romanian press published at least thirty distinct accusations of corruption and graft against officials at Metrorex, the Ministry of Transport, and Sindomet, all of whom are involved in overseeing the Metro. The commitment to aggressively prosecuting corruption has been an important element of Romanian political life post-EU accession, leading to changes in leadership and vision. See Mendelski, "15 Years."

Chapter 3

1. For Washington, D.C., see Kern, "It's a Cellars Market." For New York City, see Chen, "The Underground Apartment Market." For Toronto, see Suttor, "Basement Suites." For Vancouver, see Mendez and Quastel, "Subterranean Commodification." For London, see Burrows, Graham, and Wilson, "Bunkering Down?"

2. For Berlant, "cruel optimism" captures the impasse when the object of one's desires becomes the obstacle to one's flourishing. Berlant, *Cruel Optimism*, 1–2.

3. Anderson, *Imagined Communities*, 7.

4. In his detailed analysis of housing confiscation and restitution in Romania, Liviu Chelcea argues that restitution should be regarded as a genealogical practice, where the dispossessed recreate relations with ancestors and recalibrate relations with living kin. Chelcea, "Ancestors, Domestic Groups."

5. "Average Net Income."

6. On the potentially hazardous consequences of being beneath legal regulation, see Chapter 7 of this book.

7. The threat of being cast out to the periphery and down the social ladder resonates with James Ferguson's discussion of disconnection as an experience of abjection. "Disconnection, like abjection," Ferguson writes, "implies an active relation, and the state of having been disconnected requires to be understood as the product of specific structures and processes of disconnection." Ferguson, *Expectations of Modernity*, 238. Just as Ferguson argues that being unconnected is not the same as being disconnected, being priced out of the respectable part of town is a distinct indignity from being excluded from it.

8. The incorporation of basement apartments in new residential buildings is common in other tight residential markets. As the *New York Times* reported while this research unfolded, "In an analysis of Department of Buildings records in several Manhattan neighborhoods, eighty of 377 recent projects, or 21 percent, put residential space on below-grade floors." This included buildings on the Upper East Side, the Upper West Side, in Midtown, Chelsea, and downtown neighborhoods. Alongside these units, the same report notes, then Mayor de Blasio also backed a pilot program to modify the building code to allow for legal basement and cellar apartments in order to create more affordable housing. Chen, "The Underground Apartment Market." See also City of New York Press Office, "Pilot Basement Conversion Program." The phenomenon has taken on a distinctly upscale form in London. See Wainwright, "Billionaires' Basements"; Wainwright, "Millionaire Mega-Basements."

9. Pierre Bourdieu conceptualized class competition along a horizontal axis, or field, where actors maneuvered for position. See Bourdieu, *Distinction*, 226. An eye to basement apartments, as much as penthouses, brings into focus the vertical axis of class positioning, revealing how actors can maneuver beneath the city in the name of upward social mobility.

10. The underground becomes "productive" to users like Mădălina in that it affords an otherwise unavailable space for projects of self-cultivation as "autonomous," "respectable," and "stylish." See Miller and Rose, "Mobilizing the Consumer," 3; also Deleuze, *Foucault*, 103.

11. Bronislaw Malinowski famously wrote of the need to mitigate his informants' irritation with his constant questioning by offering donations of tobacco. Malinowski, *Argonauts*, 8. Such small gifts for informants remain a consistent line item in research budgets.

12. "Whenever I consider the relations between country and city, and between birth and learning," Raymond Williams writes, "I find this history active and continuous: the relations

are not only of ideas and experiences, but of rent and interest, of situation and power; a wider system." Williams, *The Country and the City*, 7. As Williams argues, interests in social positioning and hierarchy underlie the contrasting of the country from the city. The distinction between the two poles is all the more significant for Ştefan and Viva who are positioned at the penumbra of the city, physically, socially, and also to a certain extent historically as second-generation residents of the capital.

13. Iulia's visceral reaction to living in an unrenovated apartment from a socialist-era block is a widespread and well-documented feeling. Writing about nearby Budapest, Krisztina Fehérváry details how Hungary's new middle classes coded socialist-era home interiors as "abnormal" and spent luxurious sums to renovate their apartments to have highly stylized home interiors that they perceived as "normal." Fehérváry demonstrates that the discourse of the normal is indicative of middle-class aspirations, claiming "European" status by evaluating their own standards of living in comparison to imagined Western ones. Fehérváry, *Politics in Color and Concrete*.

Chapter 4

1. Racu, "Cât Costă"; Stoian, "Proprietarii Barurilor."

2. A number of global media outlets have attested to Bucharest's emerging status as Europe's new party destination, vying to take the place of Prague and Berlin. See, for example, Baker, "Bucharest's New Old City"; "VICE Guide"; Miller, "Young College Graduates"; Gillet, "New Life"; and Miller, *101 Places*.

3. For a detailed cultural history of Club A, see Ionescu, *CLUB A*.

4. Elsewhere I have written at greater length about socialist-era efforts to remake Bucharest's historic center. O'Neill, "Political Agency of Cityscapes."

5. "Reabiltarea."

6. As Ema Grama notes, the municipality believed the Old Town district's historic architecture could stand as proof of Bucharest's European history. However, given its state of disrepair, the district required significant capital investment to pave its streets and to refurbish its infrastructure. Grama, *Socialist Heritage*, 175.

7. "Reabiltarea."

8. On the eviction of Roma squatters, see Lancione, "Embodied Urban Precarity." For the rapid commercialization of central Bucharest, see Chelcea, Popescu, and Cristea, "Who Are the Gentrifiers," 124.

9. Baker, "Bucharest's New Old City."

10. Segmentarity was a term initially used by ethnologists to describe political systems linking lineage to territory in societies lacking a fixed, central state. See Dumont, "Preface"; Dresch, "Segmentation." Deleuze and Guattari argue, however, that segmentarity is central to the organization of modern society, noting that "the modern political system is a global whole, unified and unifying, but is so because it implies a constellation of juxtaposed, imbricated, ordered subsystems." Deleuze and Guattari, *A Thousand Plateaus*, 210. Old Town's nighttime economy, as this chapter details, proliferated along a segmentary logic.

11. Sorin spoke of this strategy as self-evident in its economic rationality. The targeted investment into basements and cellars in the name of yielding quick returns, rather than investing in the stability of the building as a whole, reflects the temporal logic of advanced capitalism. In her ethnography of Wall Street, Karen Ho details investor strategies that necessarily prioritize short-term profits at the expense of investing in projects that bend toward

long-term social stability. Ho, *Liquidated*, 164–65. The pursuit of short-term investments and quick returns, Emily Martin further argues, has led to a rethinking of the "person" as nation states and corporations cede responsibility for long-term social organization to the interests of global flows of capital in ways that can be jarring to middle-class subjects. Martin, "Flexible Survivors," 514. As this chapter makes clear, Bucharest's middle classes ultimately bear the negative consequences of these short-term investment strategies.

12. Emil Barbu (Mac) Popescu was a student of architecture in the 1960s and founded Club A as a leisure space for fellow students. Popescu went on to earn his doctorate in architecture and served as a professor and, ultimately, President of the Universitatea de Arhitectură și Urbanism "Ion Mincu" (UAUIM) from 2008 to 2012. His keen insight into the history and spatial trends of Bucharest's basements and cellars was most helpful.

13. Others have also noted the distinctly targeted renovation of Old Town's basements and cellars. See Bürkner and Totelecan, "Assemblages of Urban Leisure Culture."

14. For the Old Royal Court, see the Introduction of this book. For histories of basement jazz clubs and speakeasys, see Burke, *Come in and Hear the Truth* and Heap, *Slumming*. For London, see Haslam, *Life After Dark*. For Berlin's cabarets, see Jelavich, *Berlin Cabaret*. For a history of the ratskeller, see Meyer, *In the Good Ratskeller of Bremen*. The hashtag #UndergroundBar on Instagram attests to the genre's enduring, worldwide appeal.

15. The townhouses of Bucharest's merchants and craftsmen were a small-scale version of an early eighteenth-century squire's court. Vintilă describes a series of examples, with country-style homes built atop a raised first floor to accommodate arched cellars below, a porch, and a ground floor divided into smaller rooms for entertaining. Vintilă, *Bucureștiul subpământean*, 80–88.

16. See Vintilă, *Bucureștiul subpământean*, 84.

17. "Facadism" refers to the preservation of historic facades, or to the creation of facsimiles, to serve as decorations for new buildings constructed immediately behind them. Richards, *Facadism*. As Fredric Jameson argues, facadism produces a disconnect between a building's exterior presentation and its interior organization and use that is characteristic of postmodernism. Jameson, *Postmodernism*, 115.

18. Lewis Mumford distinguished underground from aboveground space as a site where the organic is displaced by the inorganic, where the environment is manufactured by modern engineering rather than an effect of natural processes. Mumford, *Technics and Civilization*, 70. In a similar vein, Michel Foucault referred to the production of parallel and juxtaposed temporalities, or heterochronies, that can be opened up in specific spaces. Foucault, "Of Other Spaces," 26. Foucault noted two kinds of heterochronies. The first reflect the indefinite accumulation of time, as found in museums and in libraries; the second and opposite kind pertain to time in its most fleeting, transitory, precarious aspect, such as the festival or, in this instance, the endless night found inside of Bucharest's basement and cellar clubs.

19. "The commodity prevails over everything. (Social) space and (social) time, dominated by exchanges, become the time and space of markets," observed Henri Lefebvre in *Rhythmanalysis*, 16. For Lefebvre, "commodity time" or "commodity rhythm" referred to modernity's break with time as cyclical and organic, marked by the Earth's daily rotations and yearly revolutions, in favor of the linear time of the mechanical production of commodities. In Old Town's basement clubs, the economy, rather than the cosmos, becomes the central organizer of society.

20. Michel Foucault spoke of the late twentieth century as an "epoch of simultaneity." "We are in the epoch of juxtaposition, the epoch of the near and far, of the side-by-side, or the

dispersed. We are at a moment . . . when our experience of the world is less that of a long life developing through time than that of a network that connects points and intersects with its own skein." Foucault, "Of Other Spaces," 22. This analysis extends Foucault's horizontal perspective to explore the vertical dimensions of simultaneity, tracing how different projects become stacked one atop of the other.

21. Collective effervescence refers to an energy or sentiment shared among a group of participants. The shared sense of excitement within individuals, Émile Durkheim argued, fosters a sense of social unity between them. Durkheim, *Rules of Sociological Method*, 171.

22. "#Colectiv."

23. Preda et al., "Incendiu fără precedent"; Pop, "Incendiu în Clubul Colectiv."

24. "#Colectiv."

25. Doroftei, "Incendiu în Colectiv."

26. "Incendiu în Clubul Colectiv"; also, Dobrescu, "Primul termen."

27. Iolu, "O traumă națională."

28. Preda and Vancu, "Reacții oficiale."

29. Elsewhere I have explored the political activism forged by this tragedy; see O'Neill, "Corruption Kills." Adrian Deoancă poignantly characterized the mass protests following the Club Colectiv fire as an expression of "middle class virtuousness, grounded in an ethics of personal responsibility, manifested not only through calls to civic engagement and support for technocratic anti-politics, but also through demands for moral and physical cleanliness." See Deoancă, "Class Politics." In Chapter 7 of this book, I will return to the Colectiv fire while discussing in greater depth the dangers, both real and imagined, of extending urban life underground.

Chapter 5

1. Argenbright, "Avtomobilshchina." Sherouse, "Where the Sidewalk Ends."

2. See Melly, *Bottleneck*, 7; and Truitt, "On the Back of a Motorbike," 12.

3. Jastrzab, "Cars as Favors in People's Poland," 36.

4. Peteri, "Alternative Modernity?," 47.

5. As the municipality's parking strategy notes, older residential zones in the city center were not designed with garages or other parking facilities. PMB, "Strategia de parcare," 15. Residential blocks and office buildings built before the 1990s offer only a handful of parking places, usually behind the buildings. Moga, "Mall-urile."

6. Moga, "Mall-urile"; "3.000 de mașini abandonate."

7. "3.000 de mașini abandonate."

8. In the longer-established consumer societies of Western Europe and North America, city planners have typically produced at least four parking spaces for every car in circulation to accommodate middle-class automobility. Henderson, "Spaces of Parking," 70–71. By some estimates, U.S. cities have an average of about eight parking spaces for each car. Ben-Joseph, *ReThinking a Lot*, 13. The global demand for car parking has been estimated at 19 billion parking spaces, or a parking lot the size of France. Shoup, "High Cost," 8.

9. "Mii de mașini abandonate."

10. On the estimated number of abandoned cars, see Iancu, "Peste 200.000 de mașini vechi." A vehicle qualifies as abandoned if it has been left unattended in a public space for at least one year. "Atentie." On the repurposing of derelict cars as storage space, see "Mii de masini vechi."

11. Synergetics Corporation et al., "Plan integral," 6.

12. At the time, demographers estimated that automobility would increase by 93 percent from 2007 to 2027. Planners based their estimates on the projected increase in Gross Domestic Product per capita. Consiliul General al Municipiului București, "Master planul," 4.

13. An objective of Bucharest's Integrated Urban Development Plan was to streamline traffic in the city center by creating more and smaller parking spaces located away from public space. These objectives aligned with the traffic strategies of other EU capitals. Synergetics Corporation et al., "Plan integrat," 136.

14. With the development of underground parking, planners intended to eliminate street parking wherever possible. PMB, "Strategia de parcare," 12. For information about the proposed location, cost, and size of each specific underground parking garage, see PMB, 49. While city planners built more spaces, some planners questioned the ethics of building new parking. From this critical viewpoint, the production of new parking lots only further encourages personal automobility, spiking the demand for parking ever more. see Ben-Joseph, *ReThinking a Lot*; Shoup, "High Cost"; Henderson, "Spaces of Parking"; Sherouse, "Where the Sidewalk Ends," 448.

15. With space in the city center already at a premium, locating parking garages underground, the document reasoned, enabled the city surface to be developed toward more profitable ends. PMB, "Strategia de parcare," 13. In another document, planners argued that, to preserve Bucharest's European character, the surface of the city center should be reserved for commercial uses rather than transit. As the "Integrated Development Plan for Central Bucharest" reads, "The streets and boulevards in the central area must facilitate the support of urban life, not primarily transit. The arteries in the center must be treated as streets of a local character that support commercial, cultural and leisure activities adapted to a European urban center." Synergetics Corporation et al., "Plan integrat," 147.

16. The aesthetic concern also appears in planning documents: "multi-storied general parking has to be constructed underground due to its visual impact and because of the lack of space for constructing a parking garage building [aboveground]." PMB, "Strategia de parcare," 13.

17. Consiliul General al Municipiului București, "Master planul," 6.

18. PMB, "Strategia de parcare," 23.

19. PMB, 24.

20. PMB, 32.

21. Bucharest is not unique in its Parisian fantasies. For many, nineteenth-century Paris represents modernity's aesthetic ideal. See Baudrillard, "Modernity," 68. As romantically captured in Walter Benjamin's *Arcades Project*: "Thus appear the arcades-first entry in the field of iron construction; thus appear the world exhibitions, whose link to the entertainment industry is significant. Also included in this order of phenomena is the experience of the flaneur, who abandons himself to the phantasmagoria of the marketplace." Benjamin, *Arcades Project*, 14–15. Unlike contemporary forms of consumer culture that privatize the individual, the phantasmagoria of nineteenth-century Paris celebrated the public and inclusive potential of exploring the city's enchantments.

22. Radu, "Înființarea."

23. Taking inspiration from theater studies, Erving Goffman made a fundamental sociological distinction between the "frontstage," where social performances occur, and the "backstage" where actors get themselves into character. Goffman, *Presentation of Self*, 111–12.

A dominant metaphor in the social sciences, this chapter shifts perspectives from the horizontal axis of front and back to bring into view the work performed along the vertical axis of substage and stage. In the nineteenth century, for example, a theater's substage consisted of three levels, totaling nine meters deep. Like the backstage, the substage was situated outside of the audience's view. However, the substage was not populated by actors preparing for their role on stage but instead was staffed by crewmen whose work was integral to scene setting. Workers in the substage manipulated cranks, gears, and pulleys to set the scene and to execute sound effects. Pendle and Wilkins, "Paradise Found," 178, 190. The substage is infrastructural as well as a lived space where the work of generating atmosphere is performed. See Fitzgerald, *World Behind the Scenes*, 47. This chapter takes parking garages as inhabited, infrastructural spaces oriented toward staging and scene-setting.

24. Interparking opened its first multistory car park in Brussels in 1957. At the time of this writing, its operations had expanded to 898 facilities spread across eight European countries, generating €437.8 million annually in revenue. Interparking Group, "Activity Report 2017," 2.

25. Foucault wrote of normalization as a productive form of political power based upon inclusion, qualification, and correction. "The norm's function is not to exclude and reject," Foucault began. "Rather, it is always linked to a positive technique of intervention and transformation, to a sort of normative project." Foucault, *Abnormal*, 50. As others have noted, citizens of formerly socialist European countries have been regularly subjected to the normalizing gaze of West European consumerism (Greenberg, "On the Road to Normal"; Fehérváry, *Politics in Color and Concrete*; Patico and Caldwell, "Consumers Exiting Socialism") and liberal democracy (Verdery, *What Was Socialism*; Dunn, "Standards and Person-Making"). As the case of Interparking makes clear, normalizing power is not necessarily juridical or medical but may instead be diffused across a multiplicity of institutions, including parking.

26. Interparking Group, "Piața Universității (Bucharest)." The garage opened to much public fanfare. Marin, "Sorin Oprescu." For the total amount invested into the University Square garage, see Ofițeru, "Parcarea subterană de la Universitate."

27. Interparking Group, "Activity Report 2012," 10.

28. "Teatrul Național"; Răduță, "Parcarea de la Intercontinental."

29. Interparking's annual report boasts of "comfortable car parks and leading-edge technological equipment," adding that its University Square garages offer nothing short of the "digitization of the Interparking experience." Interparking Group, "Activity Report 2017," 6. Local news coverage of the garage's opening followed similar talking points: "At the same time, parking has a dispatch center in which all monitoring and alarm systems, such as fire detection, theft alarm, car management system, and an alarm to overcome the CO_2 concentration are centralized. . . . Within the underground car park there are sockets where electric cars can be charged, as well as special places for people with disabilities that are located near the lift. At ground level there is a civil defense shelter." Alexe, "Parcarea subterană de la Universitate."

30. The University Square Garage received a 2017 European Standard Parking Award (ESPA). Interparking Group, "Activity Report 2017," 13.

31. At opening in 2012, parking rates at the University Square garage were two and a half lei per hour during the day and one leu per hour on nights and weekends. Monthly subscriptions with unlimited parking were 499 lei for cars. Gheorghe, "Parcarea subterană." By 2019, rates had risen to five lei an hour, 488 lei for an unlimited monthly pass, 101 lei for a monthly night pass, and 341 lei for a monthly day pass. See Interparking Group, "Piața Universității (Bucharest)."

32. Cozmei, "Ce probleme."

33. Ignat, "Zeci de șoferi."

34. Newspapers closely covered empty garages. See, for example, "De ce parcarea de la Străulești"; "Metrorex"; Rudnițchi, "De ce parcarea."

35. See Chelcea and Iancu, "Anthropology of Parking," 65.

36. Epure, "Am umblat."

37. In 2010, the municipality of Bucharest issued Law nr. 155, which reorganized what was formally known as the Community Police (Poliția Comunitară) into the newly created "Local Police." "Atribuții extinse."

38. News accounts provide an additional gloss on Cezar's account, describing how police officers photograph the registration number of illegally parked cars to mail their owners 280-lei fines. Journalists credited these technologies with making ticketing more efficient and multiplying the number of fines issued. See Ignat, "Zeci de șoferi."

39. Dennis Rodgers describes "disembedding" as a strategy of spatial differentiation in cities, one in which social, cultural, and economic relations become detached from their localized contexts. Rodgers, "'Disembedding' the City," 123.

40. "Rezultatele concursului."

41. "Amenajarea Pieței Universității."

42. "Rezultatele concursului."

43. Gernot Böhme makes an important distinction between a city's image and its atmosphere. "A city's image is an expression of its self-presentation, what impression it makes," Böhme begins. By contrast, "the atmosphere of a city is precisely the way life unfolds within the city." Böhme, "Atmosphere of a City," 6–7. For Böhme, atmosphere is endemic to the daily life of a city and its neighborhoods. This is why the conscious production of an atmosphere requires dramatic interventions into the organization of space and society, such as, in this case, the excavation and development of the underground as substage.

44. "The development of public, green, event spaces improves the quality of urban space, extends urban prestige, which is an important source for attracting investments and growing the urban economy," insists the Strategic Concept for Bucharest 2035. Patron et al., "Conceptul strategic București 2035," 163. In these planning documents, aesthetics and need are intimately tied to speculations as to how the city appears at arm's length, from the perspective of foreign investors.

Chapter 6

1. Surcel, "Necropola."

2. Soficaru et al., "Date antropologice preliminare," 229.

3. Soficaru et al., "Altered Shapes, Same People," 177–78.

4. Surcel, "Necropola."

5. Mironescu, "40 de schelete umane."

6. Synergetics Corporation et al., "Plan integrat," 140.

7. Romania's first law for the protection of cultural heritage is Law No. 63/1974. It created the obligation to research any land prior to construction work, and it placed the cost of archaeological excavation onto the investor. Musteață, "Preserving Archaeological Remains in Situ," 16. On July 22, 1996, Romania signed the Council of Europe's European Convention on the Protection of the Archaeological Heritage (Revised). Romania's own Government Ordinance No. 43/2000 introduced into Romanian law the European definition of preventive archaeology.

The Romanian government formally guarantees the preservation of archaeological heritage, mandating the implementation of the methods, techniques, and practices necessary to obtain maximum information about uncovered archaeological heritage. Musteață, "Preserving Archaeological Remains in Situ," 16.

8. For a detailed account of the process and practice of stratigraphic excavation, see Caraher, "Slow Archaeology," 428–30.

9. New abilities to make microscopic recordings of strata have revealed unprecedented insights into the daily activities of past civilizations and represent a major advance in the practice of archaeological investigation. While more insightful, the heightened precision of such recordings further slows the documentation and clearing of strata. Shillito et al., "Microstratigraphy of Middens."

10. The project director for the Çatalhöyük exaction in Turkey, Ian Hodder, reflects openly upon the temporal disjuncture between archaeological analysis and the pace of development, writing: "I sometimes wonder whether modern archaeology is possible—there is such an enormous disjunction between the scientific requirements and expectations and the public (or private) purse. . . . The people with big money want so much more than microdetail—e.g., reconstructed rooms, museums, and car parks. To do that we need to move earth. But we aren't." Hodder, *Towards Reflexive Method*, 125.

11. Bear, "Time as Technique," 490. See also, Bear, "Capitalist Time," 146–47.

12. Surcel, "Necropola."

13. For a discussion of the Old Princely Court and of the subterranean tunnels running beneath historic Bucharest, see the Introduction of this book.

14. "Descoperirile arheologice."

15. Soficaru et al., "Date antropologice preliminare," 234–35.

16. "Parcare sau sit arheologic?"

17. I have discussed the scale, scope, and political significance of the redevelopment of central Bucharest in O'Neill, "Political Agency of Cityscapes."

18. For a discussion of the polyfocal formation of medieval Bucharest, see the Introduction of this book.

19. Grama, *Socialist Heritage*, 91. "To territorialize history, in socialist Romania, and in the socialist bloc in general," Grama writes, "meant to implicitly pursue a spatial centralization, one that mirrored and extended the project of creating a heavily centralized political and economic system."

20. Ceaușescu's government used a massive earthquake on March 4, 1977, which measured 7.5 on the Richter scale, as a pretext for redeveloping the city center. Although preservationists felt much of the city's historic architecture could be preserved, central planners labeled a large area of the city center as irreparably damaged by the earthquake and in a state of "imminent collapse," triggering its demolition. Giurescu, *Razing of Romania's Past*, 38.

21. "Descoperirile arheologice."

22. Kaparosa and Skayannisb, "Dealing with Context," 407

23. For New York, see Bloodworth and Ayres, *New York's Underground Art Museum*. For Moscow, see Kuznetsov et al., *Hidden Urbanism*.

24. The construction of the planned Metro hub in Istanbul's Yenikapi neighborhood gave way to a roughly four-year stratigraphic excavation of an unearthed Byzantine port, uncovering some thirty-four well-preserved shipwrecks. See Bonini Baraldi, Shoup, and Zan, "When Megaprojects Meet Archaeology," 430; Rose and Aydingün, "Under Istanbul."

25. In her work on South Africa, Lynn Meskell speaks of "making heritage pay," to capture the increasing entanglement of heritage with the rhetoric of sustainability, tourism, and improving socio-economic conditions. Meskell, *Nature of Heritage*, 38. As the World Bank and other development institutions turn to heritage as an asset to be leveraged, development priorities, rather than conservation concerns, increasingly inform heritage practices. Samuels, "Heritage Development." See also Bertacchini and Saccone, "Toward a Political Economy of World Heritage." In Bucharest, municipal and market actors looked to harness the economic potential of the city's architectural heritage at the expense of archaeological remains, which were seen as having little economic potential despite their historical value.

26. Synergetics Corporation et al., "Plan integrat," 5.

27. Patron et al., "Conceptul strategic Bucureşti 2035," 168.

28. For an account of the movement of Bucharest's churches during socialist-era development projects, see Danta, "Ceauşescu's Bucharest," 179–81.

29. See Oberlander-Tarnoveanu, "Preventive Archaeological Research," 168; Măgureanu and Măgureanu, "Preventive Archaeology," 259.

30. A similar tension between private, contract archaeology and academic archaeology is both long established and clearly documented in the U.S. and U.K. contexts. See Green and Doershuk, "Cultural Resource Management and American Archaeology."

31. See Oberlander-Tarnoveanu, "Preventive Archaeological Research," 172.

32. As Măgureanu and Măgureanu argue, "local authorities need money from taxes and a new investor means more money for the local budget . . . local authorities have neither a correct nor a complete map of archaeological sites, and therefore no interest in protecting the sites—even if they perhaps understand the importance of doing so. Thus, they allow more and more buildings or other kinds of projects to affect archeaological sites, all in the name of developing their community." Măgureanu and Măgureanu, "Preventive Archaeology," 266.

33. Măgureanu and Măgureanu, 266.

34. "County culture departments and the Regional Commissions of Historical Monuments have no instruments to protect [archaeological sites] or provide for their preventive research," warns Sorin Nemeti. Underfunded and understaffed, these departments are unable to monitor sites and to enforce regulations. Nemeti, "Manifesto," 7.

Chapter 7

1. "Top 10."

2. Gomez, "Europe's Hottest Building Market."

3. "About SkyTower."

4. Paul Virilio argues that the invention of new technologies is also the indirect invention of their catastrophic breakdown. Virilio writes, "To invent the sailing ship or steamer is to invent the shipwreck. To invent the train is to invent the rail accident of derailment. To invent the family automobile is to produce the pile-up on the highway." Using the underground as a metaphor, Virilio argues that accidents provide an analysis of "What is beneath (substance) any knowledge." Virilio, *Original Accident*, 10. As a basic observation, Virilio continues, "If inventing the substance means indirectly inventing the accident, the more powerful and high-performance the invention, the more dramatic the accident." Virilio, 31. From Virilio's perspective, the accident reveals the limitations of knowledge about the technologies that are rapidly incorporated into modern life. He understands accidents not as a matter of individual blame but as a cumulative effect of modernity's technological reliance.

Quoting Hannah Arendt, Virilio posits that "progress and disaster are two sides of the same coin." Virilio, 15. Following Virilio, this chapter understands the production of the underground as the production of the accidents that are endemic to the project of developing downward—in and beyond Bucharest.

5. The Millennium Business Center changed ownership over the course of its construction. The British investment fund European Convergence Property Company purchased the building in 2006. "Millennium Business Center." The building was then quickly sold again to the German investment fund DEGI. Zamfir, "Ghost Story."

6. See "Millennium Business Center."

7. Sîrbu, "Fantoma."

8. Zamfir, "Ghost Story."

9. Ivan, "Interviu Noica."

10. Local civil engineers regularly noted Bucharest's geological and architectural difficulties. As one summarized, "Because of the . . . numerous floods, fires, earthquakes, and the tragic communist era of demolitions, [the development of large building blocks,] and chaotic public network development, the city can be considered a real trap. . . . The complex architectural and urban city context . . . increase the difficulty of designing underground structures in the urban environment and implicitly increase also the difficulty of positioning and executing geotechnical investigations in situ." Toma et al., "Necessary Geological and Geotechnical Information," 714.

11. "Petronas Twin Tower 1."

12. Sergiu's acknowledgement of the space between the soil's calculated and predicted behavior and its actual behavior calls to mind Paul Virilio's description of accidents as a kind of "sabotage of prospective intelligence." Virilio, *Original Accident*, 22. To take into account the known presence of unknown variables, developers are legally required to closely monitor soil pressure during and after construction. Quickly catching the soil's unpredicted shifts helps to minimize the negative impact of excavation on a given project and its surroundings.

13. Geofri, a civil engineer whose firm executes road design and repair for the municipality, echoed Sergiu and Luca's concerns. "Whenever there is an excavation near the road," Geofri explained, "any kind of vibration from construction, any extraction of land, any movement of the terrain can lead to cracks and potholes." From Geofri's perspective, the proximity of construction poses as great a concern as its depth.

14. The skirting of safety regulations is not particular to Bucharest. The development of soaring skylines in other cities also outpaces the capacity of governments to regulate and the private sector's ability or desire to self-regulate. See Chen, "New Supertalls."

15. The rate and speed of industrial production, Virilio argues, has changed the character of the accident from one of chance to serial production: "In the course of the twentieth century, the accident became a heavy industry." Virilio, *Original Accident*, 12. As construction outpaces regulation, the accident becomes an inevitable part of urban life.

16. "Euro Tower: Bucharest, Romania." Also, "Euro Tower."

17. Ivanov, "Prețul plătit."

18. "Lacul Circului"; Barbu, "Lacul a scăzut."

19. "Lacul Circului"; "Un lac din București." City Hall, for its part, deflected responsibility for Circus Lake's changing water level away from development. In contrast to the opinion of environmentalists, City Hall issued a report attributing the drop in water level to decreased

precipitation. Ivanov, "Prețul plătit." Experts in subterranean water systems, however, asserted that building foundations were partially responsible for drops in the water level of Circus Lake. Gogu et al., "Urban Groundwater," 7.

20. Ivanov, "Prețul plătit."

21. "A society that unthinkingly privileges the present to the detriment of past and future," Virilio writes, "also privileges accidents. Since, at every moment, everything happens, most often unexpectedly, a civilization that implements immediacy, ubiquity, and instantaneity, stages accidents and disasters." Virilio, *Original Accident*, 23. To this list, Radu added that a society that privileges individual responsibility over and above forms of social responsibility also privileges accidents.

22. For a fuller discussion on the impact of fluctuating groundwater levels, see Gogu et al., "Urban Hydrogeology Studies," 136.

23. ADRBI, "Planul," 94.

24. "1,3 miliarde litri/lună."

25. Răduță, "Groapă mare."

26. Răduță, "A fost reparată avaria."

27. Răduță, "S-a rupt asfaltul."

28. Cozmei, "Ce s-a întâmplat la metrou."; "Pericol de surpare."; Serea, "E haos." For a discussion of the site's complexity, see Bitetti et al., "Construction Methods."

29. Outside observers confirm as much. The World Economic Forum recently ranked Romania's road infrastructure (road quality and comprehensiveness) 120th out of 137 countries, the lowest score in the EU. World Economic Forum, "Quality of Roads." Romania's infrastructure has ranked last since it joined the EU in 2007. Timu and Vilcu, "Facelift."

30. As Geofri noted, the destabilization of the city surface because of subterranean water (mis)management has been long documented in New York. In 1906, for example, crews constructing underground train tunnels in Manhattan excavated through a buried stream. Water leakage caused the surrounding soil to settle. Building foundations sank and a truck plunged through the pavement on 33rd Street after the subsurface soil had washed away. Solis, *New York Underground*, 106.

31. Dumitrescu, "Ancheta incendiului."

32. Underground structures have been considered relatively safe from earthquake effects compared to structures above the ground. Experts had assumed that underground structures followed the deformation of the ground during an earthquake. Yoshida and Nakamura, "Damage to Daikai," 287. However, more recent research has detailed how underground structures respond differently to seismic events relative to aboveground structures. The main strain on underground structures, for example, is not obviously related to the magnitude of earthquake acceleration, but instead is closely related to the strain or deformation of the surrounding rock and soil. For the upper structure, seismic acceleration is an important factor affecting its dynamic response. See, Liao, "Overview."

33. Lanzano, Bilotta, and G. Russo, "Tunnels under Seismic Loading," 65.

34. The Hyogoken-Nanbu earthquake of January 17, 1995, for example, led to the complete collapse of the Daikai subway station, west of downtown Kobe City, Japan. The floor of the station was subjected to a strong horizontal load from the adjacent subsoil which caused deformation of the station's box frame. The station's columns bent and sheared, causing the ceiling to collapse underneath the weight of the overburdened soil. Yoshida and Nakamura,

"Damage to Daikai," 295. Substantial damage to subway tunnels was also observed along the Chengdu Metro Line in China following the 2008 Wenchuan earthquake. Liao, "Overview"; Juan, "Damages of Metro Structures."

35. Engineers have recently pointed to the complex interactions that can occur between surface structures, soil, and underground subway structures. Important factors mediating the impact of these interactions include the horizontal distance between the aboveground buildings and subway station, the height of the buildings, the burial depth of the subway station, and the number of surface buildings. Miao et al., "Seismic Response," 3; Zhuang et al., "Seismic Performance Levels."

36. Myers, "Please Save Us!" In a similar vein, in New York City, record-shattering rainfall from Hurricane Ida in 2021 drowned eleven people living inside illegally converted basement apartments not unlike those found in central Bucharest (see Chapter 3). These New York apartments provide low-cost housing opportunities within the city as well as additional income for their typically middle-class landlords living in the property above. Ferré-Sadurní et al., "Death Traps."

Chapter 8

1. Băltăreţu, "Biblioteca Digitala revine."

2. McCann Bucharest received a Silver Cannes Lion and a Bronze Lion for its Digital Library campaign. See Blănaru, "McCann Erickson Bucharest." Copycat campaigns then took shape in London (Onwuemezi, "Shakespeare 'Digital Library Wallpaper'"), New York (Bridger, "Empty Libraries"), and Beijing (Bănilă, "Operatorul Metroului din Beijing").

3. Crăciun, "Elevii din 300 de licee."

4. The historian Patrick Joyce argues that the idea of the free library was central to the creation of a new sort of public. "The library created the liberal citizen in large measure through the fostering of self-help and self-culture, both of which involved self-knowledge. In fact, this was a constant operation on the self that directly represented the moral self-address that has been seen to be intrinsic to a liberal ethics of governance." Joyce, *Rule of Freedom*, 130. This chapter explores the vertical formation of a public and what it means to engage in the public life of Metro stations, trains, tunnels, and vestibules.

5. Mumford, *Culture of Cities*, 239.

6. Martin Heidegger drew upon the experience of waiting for a train to explore the two structural elements behind the experience of slowed time. The first is a temporal misalignment between the passenger waiting on the platform and the train that has yet to arrive, and which holds the passenger in limbo; the second is a spatial relationship between the passengers and the "tasteless station," which leaves the passenger feeling empty. Heidegger, *The Fundamental Concepts of Metaphysics*, 86, 93. Heidegger's analysis of being held in limbo and left empty resonates with passenger descriptions of boredom and feeling stuck while riding the aesthetically uninspired Bucharest Metro.

7. Teresa Caldeira neatly summarizes the modern ideal of urban public space as follows: "Although there are various and sometimes contradictory accounts of modernity in Western cities, the modern experience of urban public life is widely held to include the primacy and openness of streets; free circulation; the impersonal and anonymous encounters of pedestrians; spontaneous public enjoyment and congregation in streets and squares; and the presence of people from different social backgrounds strolling and gazing at others, looking at store windows, shopping, sitting in cafes, joining political demonstrations, appropriating the streets

for their festivals and celebrations, and using spaces especially designed for the entertainment of the masses (promenades, parks, stadiums, exhibition spaces). These are elements associated with modern life in capitalist cities at least since the remodeling of Paris by Baron Haussmann in the second half of the nineteenth century." Caldeira, *City of Walls*, 299.

8. Jürgen Habermas pointed to the coffee house as a critical institution in the shaping of a liberal public in Great Britain, France, and Germany. In contrast to the Court as a residence of secluded royalty, Habermas wrote of coffee houses as "centers of criticism—literary at first, then also political—in which began to emerge, between aristocratic society and bourgeois intellectuals, a certain parity of the educated." Habermas, *Structural Transformation*, 32. Inside the coffee house, the authority of the better argument could assert itself against that of social hierarchy, helping to instill a sense of "common humanity." Habermas, 36.

9. As Benedict Anderson argues, "the nineteenth century was, in Europe and its immediate peripheries, a golden age of vernacularizing lexicographers, grammarians, philologists, and literatures. The energetic activities of these professional intellectuals were central to the shaping of nineteenth-century European nationalism." Anderson, *Imagined Communities*, 73. Their efforts, Anderson notes, were collected, circulated, and consumed within the great libraries of Europe, particularly within university libraries. The sociologist Eric Klinenberg affirms the library's enduring place in public life, arguing that "social infrastructure provides the setting and context for social participation, and the library is among the most critical forms of social infrastructure we have." Klinenberg, *Palaces for the People*, 32.

10. Taking the example of the mine as his case in point, Lewis Mumford wrote: "Within the subterranean rock, there is no life, not even bacteria or protozoa. . . . Except for the crystalline formations, the face of the mine is shapeless: no friendly trees and beasts and clouds greet the eye." Mumford, *Technics and Civilization*, 69. Life inside of such "manufactured environment[s]," Mumford bemoaned, lacked stimulation. They were simply unhospitable places.

11. Take for example the figure of the flâneur and the pleasures and excitements of *flânerie* as chronicled by Walter Benjamin and developed by others. Benjamin, *Arcades Project*; also Buck-Morss, *Dialectics of Seeing*. Reading the city, in this tradition, presents itself as an evolving source of pleasure and excitement.

12. Benjamin, *Arcades Project*, 423–24. The fundamental experience of the flâneur, David Frisby notes, is the "sensational phenomenon of space," as articulated by "the streets and their architecture, the ostentatious architecture of mass transit (railway stations), masses of spectators (exhibition halls) and mass consumers (department stores), to which the flâneur is drawn." Frisby, *Cityscapes of Modernity*, 39. The excitement of metropolitan life presents itself through the spectacles of streets and buildings and of the masses of people coursing through them.

13. Buck-Morss, "Flaneur," 102. Similarly, Walter Benjamin fretted over the fate of street names, which lay outside of the perceptions of those stuck in the vaults of the Métro. Benjamin, *Arcades Project*, 840.

14. After the end of socialism, Krisztina Fehérváry noted, people across Eastern Europe used the idea of "normal" to refer to objects, services, and standards of living which were clearly extraordinary in their local context and yet were imagined to be part of "average" lifestyles in Western Europe. The materially richer standard of living found in Western Europe became the implicit norm, however unattainable, against which post-socialist populations judged the quality of their home life specifically and material culture more broadly. Fehérváry, "American Kitchens," 370–71.

15. "This techno-imaginary universe—of digital eras and divides—has the effect," Faye Ginsburg argues, "of reinscribing onto the world a kind of 'allochronic chronopolitics,'" one in which those without certain devices exist in a time not contemporary with our own. Ginsburg, "Rethinking the Digital Age," 291.

16. In developing his notion of "street life," Ato Quayson describes what he calls the "colonization of the sidewalk" along Oxford Street, Accra, by both commerce from shops and stores as well as informal vendors of various kinds, giving the street a distinctly effervescent character. "To walk along Oxford Street is also to be constantly invited to pause and look at things: not in the manner by which shop windows in commercial boulevards elsewhere pose various enticements for the pedestrian to stop . . . but rather by the constant barrage of vendors of all manner of goods vying to make a sale." Quayson, *Oxford Street, Accra*, 16.

17. Belzberg, *Children Underground*.

18. Iris Marion Young argues that the diversity of activities found in public space is what makes the city interesting, draws people out in public to them, and gives people pleasure and excitement. Young writes that we "take pleasure in being open to and interested in people we experience as different." Young, *Justice*, 239.

19. Jane Jacobs defines a public character as "anyone who is in frequent contact with a wide circle of people and who is sufficiently interested to make himself a public character. A public character need have no special talents or wisdom to fulfill his function. . . . His main qualification is that he *is* public, that he talks to lots of different people. In this way, news travels that is of sidewalk interest." Jacobs, *Death and Life*, 68. The informal, if not entirely self-appointed, role of public character is, Jacobs insists, essential to the structure and character of a vibrant sidewalk and to establishing a sense of community.

20. Cristescu, "Blocaj la metrou"; "Probleme la metrou."

21. Bucharest has a history of overwhelmed public transit. One account of buses from the socialist era notes that "it is a common sight to see vehicles so overloaded that passengers are clinging onto the outside of the bus." Banister, "Transport in Romania." 261. Writing in the same era, Katherine Verdery documented "trains [that] were so crowded that most people had to stand, making it impossible to use the time to read or work." Verdery, *What Was Socialism*, 47.

22. Sindomet Servcom SRL manages advertising and commercial space across the Metro system. Since 2007, Metrorex has also partnered with Euromedia Group and Spectacular OOH Printing to administer advertising. "Directorul General Metrorex."

23. I conceptualize this transformation of Victoriei Station as a form of what Deleuze and Guattari call "deterritorialization": an "uprooting" or "metamorphosis" that fundamentally transforms the cultural coordinates of a territory. Deleuze and Guattari, *A Thousand Plateaus*, 21. The cultural reworking of a territory, Brian Massumi explains, unfolds through a process of territorial "decoding," which is then followed by a "reterritorialization" that imposes new patterns of connection onto a territory and its surroundings. Massumi, *User's Guide*, 51. The DPL deterritorialized the Victoriei Station interchange and reterritorialized it as a library, at the same time transforming passengers into patrons and potential consumers.

24. Vodafone announced at the time that its 4G network covered all 41 of Romania's counties. Gabor, "Vodafone."

25. McCann Worldgroup, "Vodafone: Digital Library Wallpaper."

26. One study placed Romania last in Europe for book consumption, claiming that only 2.8 percent of Romanians read a book per month while an estimated one in five Romanians

were believed to have never read a book at all. Zachmann, "România e pe ultimul loc." Such numbers compared unfavorably to France, for example, where an estimated 90 percent of people read a book a month. Tiron, "Cea mai vândută carte." Another column expressed angst over the fact that beer consumption in Romania was rising as book sales declined further. "Românii au bani de bere."

27. Tiron, "Cea mai vândută carte."

28. In her work in Kenya, Lisa Poggiali shows how mobile phones create new scales of belonging that become foundational to self-understanding. Poggiali, "Digital Futures," 255.

29. Drawing upon the work of Lewis Mumford, Rosalind Williams describes subterranean environments as models of a completely artificial environment in which the natural environment is displaced by a technological one. Williams, *Notes on the Underground*, 4. Advances in modern engineering, Williams argues, recast the traditional underground from the site of hell to one of futurity and the technological sublime. Williams, 97.

30. For a description of the Moscow Metro, see Kuznetsov et al., *Hidden Urbanism*; for the New York Subway, see Hood, *722 Miles*; for the London Tube, see Pike, *Subterranean Cities*, 35–36.

31. "The Metro captures a new image of the future," one commentator noted after the Metro's inaugural opening. Popescu and Feraru, "Convorbire," 12. "Given the Metro will be used by tens of thousands of riders a day," the same planner continued, "the priority is the smart functioning of the stations." Popescu and Feraru, 13. Throughout the interview, the planner focused upon the contribution of the Metro to equalizing access to amenities over and above the aesthetic attributes or public life of the underground itself.

32. Popescu and Feraru, 12

33. The willingness of the Metro to allow virtually every square inch of its stations to be covered by advertising in exchange for relatively low sums of money led one executive at another advertising firm to tell me that, "In the Metro, advertising is everywhere. If you have the money, you can brand anything. It's a variation of 'guerilla marketing' in that small budgets yield big results." The executive's use of "guerilla marketing" invokes radical politics in a way that resonates with the connotations of "the underground" as a countercultural space. The wordplay, however, was fundamentally out of step with the national and transnational corporations being marketed inside the Metro as well as with the underground's middle-class demographic, which served as the audience for these branding efforts. In stark contrast to its countercultural connotations, corporate branding increasingly defines the character of underground Bucharest. For a discussion on guerrilla marketing, see Fattal, *Guerrilla Marketing*.

34. Krisztina Fehérváry tracks a similar shift in aesthetics and ideology from socialism to post-socialism within the domestic sphere. See Fehérváry, *Politics in Color and Concrete*.

35. Immersive advertisements inside the Metro regularly simulated depth with dramatic effect. A 3D graphic pushing Mountain Dew Adrenaline, as another case in point, gave the appearance that one gallery floor was a wobbly wooden footbridge suspended over molten lava. Predictably, both sides of the "footbridge" ended at vending machines where the energy drink could be purchased.

36. Michael Warner charts out three kinds of publics, noting that the boundaries between them are not always sharp. The first sense of public "is a kind of social totality. Its most common sense is that of the people in general." At the same time, Warner continues, "a public can also be a second thing: a concrete audience, a crowd witnessing itself in visible space, as with a theatrical public. Such a public also has a sense of totality, bounded by the event or by the

shared physical space." Warner's third sense of public is "the kind of public that comes into being only in relation to texts and their circulation—like the public of this essay." Warner, "Publics and Counterpublics," 49–50. Following Warner's second description, the Metro system constitutes a narrower sense of public underground in the form of its ridership. The kind of public called into being by way of a text (Warner's third sense) hinges upon a different sense of audience and circulation than that constituted by the Metro platform. This chapter is most interested in the tension between the public constituted by the platform and the one called into being by the DPL.

37. Throughout the *Arcades Project*, Walter Benjamin represented the spectacle of capitalist consumerism by using the term "phantasmagoria." Benjamin introduced the term as follows: "World exhibitions glorify the exchange value of the commodity. They create a framework in which its use value recedes into the background. They open a phantasmagoria which a person enters in order to be distracted." Benjamin, *Arcades Project*, 7. For Benjamin, the rise of mass consumerism transformed nineteenth-century Paris into a commodity spectacle whose phantasmagoria was publicly consumed.

38. "Without the idea of texts that can be picked up at different times and in different places by otherwise unrelated people, we would not imagine a public as an entity that embraces all the users of that text, whoever they might be," writes Warner. Such a public exists by virtue of its address. Warner, "Publics and Counterpublics," 51, 55.

39. McCann Worldgroup, "Case Study."

40. The New York Public Library is not a public institution in the same sense as public schools or parks, notes Tom Glynn. Public libraries were not paid for entirely by taxes, nor were they publicly governed. From their inception, libraries are "an example not only of government taking on new, more expansive roles, but also its doing so in complex ways that confounded the distinction between public and private." Glynn, *Reading Publics*, 14. To make sense of the shifting sense of public implied by public libraries, I follow Glynn's call to "trace which readers they included in or excluded from the reading public and what kinds of books public libraries collected to serve the public good." Glynn, 14.

41. Groot and Hackett, "Through the PULMAN Glass," 104; see also Davies, "PULMAN."

42. To view the Digital Public Library's current catalogue, see Vodafone, "Biblioteca Digitală."

43. Glynn, *Reading Publics*, 14–15.

44. Reeser and Spalding, "Reading Literature/Culture," 668.

45. As Pierre Bourdieu quipped, "The bourgeoisie expects from art (not to mention what it calls literature or philosophy) a reinforcement of its self-assurance." Bourdieu, *Distinction*, 293.

46. "Biblioteca Digitală Reloaded."

47. Boyarin, *The Ethnography of Reading*, 3.

48. Regarding the reading practices of Ancient Greeks and Romans, see Gavrilov, "Techniques of Reading," 56. On the reading practices of ancient Jewish culture, see Boyarin, "Placing Reading," 11.

49. In *The Confessions*, St. Augustine recounted with amazement the sight of Bishop Ambrose reading to himself. Augustine wrote, "But when [Ambrose] was reading, his eyes glided over the pages, and his heart searched out the sense, but his voice and tongue were at rest. Often times when we had come . . . we saw him thus reading to himself, and never otherwise. Saint Augustine of Hippo, *The Confessions*, VI.3. See also Engelke, "Text and Performance," 80.

50. Boyarin, "Placing Reading," 19–20.

51. For Eric Klinenberg, the public library's extensive programming, organized by a professional staff who uphold a principled commitment to openness and inclusivity, fosters social cohesion among patrons who might otherwise keep to themselves. Klinenberg, *Palaces for the People*, 36.

52. Klinenberg, 37.

53. Others have made similar points, noting how digital technologies mediate between the self and abstract concepts such as modernity, globalization, and mobility. See Mcintosh, "Mobile Phones"; Horst and Miller, *Cell Phone*; Poggiali, "Digital Futures."

54. The inclination to avoid eye contact while commuting underground is not without a history. Writing about the opening of the New York City subway, Stefan Höhne argues: "In the subway, especially, staring at others was increasingly interpreted as aggressive and rude. The cloak of isolation and anonymity that passengers drew around themselves proved delicate and thin, requiring techniques that would allow them to give their faces the expression of unmoved distance. Only by controlling one's gaze was it possible to observe social distance from others while tolerating spatial proximity." Höhne, *Riding the New York Subway*, 145. See also Simmel, "The Stranger." Rather than working to bridge and connect, as traditional public libraries above ground aimed to do, the DPL merely digitized the avoidance of others.

55. Teresa Caldeira associates walls and cars with a retreat from the modern ideal of urban public space, as both kinds of architectural and technological forms reinforce boundaries and discourage heterogeneous encounters. Caldeira, *City of Walls*, 310. Amidst efforts to move the middle classes out of their cars and into the Metro, digital technologies emerge as a primary strategy for effecting separation.

Chapter 9

1. IGSU, "Situația cu coordonatele GPS."

2. MDRAP, "Norme tehnice," 30.

3. MDRAP, 7, 10.

4. MDRAP, 16.

5. MDRAP, 18–19.

6. To ready Switzerland's sprawling system of civil protection shelters, the government distributed leaflets, exhibitions, and films instructing homeowners on how to equip their family shelters with a fourteen-day store of supplies while dedicated shelter managers tended to public shelters. Berger Ziauddin, "(De)Territorializing the Home," 681. Regarding U.S. shelters, see Masco, "Life Underground," 25. For Joseph Masco, America's bunkers were always theatrical, promoting the fantasy of security and endurance even as the Rand Corporation was tasked with attempting to calculate how long it would take before the "survivors would envy the dead." Masco, 13. For a detailed account of the material excesses of contemporary bunkers in the U.S. and beyond, see Garrett, *Bunker*.

7. For Elaine Scarry, the material practicality of these shelters is a secondary concern when thinking through their impact upon public life. As Scarry puts it, "One of two things is true. Either fallout shelters are useless, in which case neither the population nor the government should have them. Or they are useful, in which case both the population and its leaders should have them." Scarry, *Thermonuclear Monarchy*, 360.

8. Departamentul pentru Situații de Urgență (DSU), "Hartă: Adăposturi de protecție civilă."

9. Pike, "Cold War Reduction," 103.

10. "În ce locuri."

11. This is true elsewhere in Europe, as Elaine Scarry notes. Sweden currently has shelters for 81 percent of its population, Finland has 70 percent, Austria has 30 percent, and Germany has coverage for 3 percent. Scarry, *Thermonuclear Monarchy*, 359.

12. Much of the literature on the securitization of urban elites focuses upon the production of walls and gates. Caldeira, *City of Walls*; Low, *Behind the Gates*. Security features, in this literature, are private initiatives that are both conspicuous and aestheticized, turning techniques designed to separate the wealthy from the rest into markers of status. Civil protection shelters function very differently than these architectural features. Shelters are government mandated and not publicly visible, nor are they prominently marketed as an upscale feature of luxury buildings. Their designed intent, furthermore, is to shelter their users not from other citizens but from external threats. They represent a semi-public space within private buildings rather than the fortification of private space from the public beyond.

13. Elaine Scarry uses the term "equality of survival" to describe the Swiss civil protection system, which has produced enough shelter space for 114 percent of its population. She characterizes the Swiss commitment to equality of survival for its entire population as "a feat of civic and moral engineering." Scarry, *Thermonuclear Monarchy*, 354–55.

14. Household budgets are tight in Romania owing to the combination of low salaries compared to elsewhere in the EU and the demands of family life. Beyond the cost of housing, for instance, Bucharest's middle classes spend upwards of a quarter of their monthly income on extracurricular education and sports programs for their children in efforts to leave them well positioned for their own professional careers. Savu, Lipan, and Crăciun, "Preparing for a 'Good Life,'" 486. Not surprisingly, Romania's household savings rates have been negative for the last fifteen years. Commission, "Household Saving Rates," 5. Within these household budget constraints, the obligation to divert even a few hundred euros toward the maintenance of a civil protection shelter can feel overbearing.

15. Tucker, *War of Nerves*.

16. *Adevărul*, "Măsuri contra atacurilor aeriene."

17. Bucur, "Acte normativ-legislative." 59.

18. Creangă-Stoileşti, *Istoria apararii civile*, 68–70.

19. Creangă-Stoileşti, *Istoria apararii civile*, 197–99; also Bucur, "Acte normativ-legislative," 60.

20. Bucur, "Acte normativ-legislative," 60; also Creangă-Stoileşti, *Istoria apararii civile*, 211–12.

21. Creangă-Stoileşti, *Istoria apararii civile*, 224–26.

22. Truxal, "Bombing the Romanian Rail Network," 19.

23. Quoted in Schaffer, *Wings of Judgment*, 56.

24. Tanaka and Young, *Bombing Civilians*, 37; Truxal, "Bombing the Romanian Rail Network."

25. Arhivele Naţionale ale României, "Serviciul arhitecturii şi oficiul tehnic: 21/1944."

26. Marea Adunare Naţională, Legea nr. 2/1978.

27. Marea Adunare Naţională, Legea nr. 2/1978. Article 22.

28. Fully equipped bunkers, connected by a network of tunnels passable by car, exist beneath each of Romania's main government buildings. "Reţeaua de buncăre antiatomice." These bunkers and tunnels are closed to the public, and limited and often contradictory informa-

tion about them has been made available. The website for the Palace of Parliament, for example, advertises an "Underground Tour" that takes guests through a portion of the palace's initial two basement levels (upward of seven are rumored to exist). When I attempted to book the underground tour over the phone, and again in person, the ticketing agents denied the existence not only of such a tour but of the basement levels themselves. Bewildered, I retold the incident to a Romanian friend later that night who laughingly assured me that the tunnels are real because they had been prominently featured on an episode of the globally popular British television show *Top Gear*. Churchward, Klein, and Trevor, "Top Gear: Episode 14.1 (Romania)."

29. The extent of the Spring Palace's underground extension gives the building an unusually "elastic" quality. Verdery, *Vanishing Hectare*. Reports on the number of the palace's rooms ranges from thirty rooms totaling five thousand square meters (Karasz, "Ceausescu's Villa") to eighty rooms and four thousand square meters ("Romanians Flock to Ceausescu's Palace"; "Ceausescu's Nuclear Bunker") to as many as one hundred and fifty rooms (Turp-Balazs, "Bucharest's Spring Palace").

30. Parlamentul României, Hotărârea nr. 560/2005.

31. Parlamentul României, Lege Nr. 481/2004. See article 45.

32. Parlamentul României, Lege Nr. 481/2004. See article 45.

33. Parlamentul României, Legea protecției civile nr. 106/1996.

34. A space may be empty, but an empty space is neither inactive nor without utility. As others have argued, "empty space is a part of reality, including organizational reality." Kociatkiewicz and Kostera, "Anthropology of Empty Spaces," 47. Although Bucharest's bunkers sit largely empty and unused day to day, they are very much a part of the organization of public life, fully incorporating some while leaving others fully exposed.

35. Bennett, "Grubbing out the Führerbunker."

36. Solomon, "Swiss Weigh Future Role."

37. As Bennett argues, the Cold War didn't follow the spatial logics of previous wars by maintaining distinct battlegrounds and home fronts. Instead, "the Cold War was everywhere and nowhere, for this was (in North Western Europe at least) a war-in-waiting, a tense existential standoff materialized in bunker-building spread out across the whole of territories rather at borders." Bennett, "Bunker's After-Life," 3. Once decommissioned, bunkers made possible new terrain for commercialization beneath Europe's highly developed centers of industry, government, and culture.

38. Elborough and Horsfield, *Atlas of Improbable Places*, 212–15.

39. Lowe and Joel, *Remembering the Cold War*, 54–55. In England, atomic shelters have also been packaged as museums geared toward heritage tourism as well as sites for live, military-themed role play. Bowers and Booth, "Preserving and Managing," 201. Bunker 51, located in East London, for example, is a Cold War–era nuclear bunker that has been transformed into a paintball and LaserTag studio where parties lay siege to a missile silo and prison complex or secure a loading bay from a zombie apocalypse. Bunker 51 has also found a niche within East London's expanding film production industry, providing a secure, industrial backdrop for filming action sequences and music videos. See "Bunker 51."

40. Elborough and Horsfield, *Atlas of Improbable Places*, 212–15. In Berlin, bunkers have been similarly retrofitted as cultural venues. The Reichsbahnbunker, for example, was built by forced labor during the Third Reich to serve as an air raid shelter for upward of three thousand people (Nabas, "Uncomfortable Heritage," 65). Immediately after German reunification,

the bunker functioned as a multi-floor techno club venue whose intensity of musical style and volume played upon the bunker's brutalist architecture. Wilson, "Sounds from the Bunker"; Schofield and Rellensmann, "Underground Heritage," 128. In 2003, the art collector Christian Boros transformed the bunker into a museum to publicly showcase his private collection. Bucknell, "Nazi-Era Bunker"; Morgan, "Bunker Conversion," 1341–43. Such creative repurposing of demilitarized bunkers, others have argued, has enabled Berliners to position their city as an ultra-modern capital and an epicenter for the production of "cool" and "hip." Schofield and Rellensmann, "Underground Heritage."

41. Reilly, "See out the Apocalypse."

42. "La Claustra."

43. A tour of the hotel is available on YouTube. Rainer, "La Claustra." See also Nalewicki, "Switzerland's Historic Bunkers."

44. "La Claustra"; Taylor, "Future-Proof."

45. The Swiss Null Stern hotel, which opened in the public bunker of Teufen in 2009, illustrates a dramatic contrast. Also a repurposed nuclear fallout shelter, the Null Stern is as much art installation as a hotel. It retained the bunker's internal structure and design, adding only minor amenities like high-end linens and Swiss chocolates to offer a juxtaposition with the bunker's concrete walls and exposed ductwork. Berger Ziauddin, "(De)Territorializing the Home," 687. The hotel's creators described the Null Stern as "a place where people can think about their surroundings." McKinlay, "Switzerland's Null Stern Hotel." Meanwhile, others have documented how government-mandated bunkers in the homes of Swiss families have been quietly converted into wine cellars, storage rooms, gyms, and band rooms, for example. Berger Ziauddin, "(De)Territorializing the Home," 686.

46. Starosielski, *The Undersea Network.*

47. The Bunker, "The Bunker."

48. BDO, "Cyberfort Group."

49. "Mount10."

50. Daly, "Nuclear Bunker."

51. Garrett, *Bunker*, 5.

52. For Foucault, biopolitics encapsulates a shift in the exercise of power, from a sovereign right to take life to a governmental power that increasingly exercises "the right to intervene to make live, or once power begins to intervene mainly at this level in order to improve life by eliminating accidents, the random element, and deficiencies, death becomes, insofar as it is the end of life, the term, the limit, or the end of power too." Foucault, *Society Must Be Defended*, 248.

53. According to DSU's spreadsheet, there are 139 shelters serving the wealthy north in Sector 1; 338 shelters in Sector 2 (northeast); 102 shelters in Sector 3 to the east; 159 shelters in Sector 4 to the south; only fifty-six shelters in Sector 5 covering the southwest of Bucharest; and 289 shelters in Sector 6 to the west. IGSU, "Situația cu coordonatele GPS."

54. Bare life, as philosopher Giorgio Agamben describes it, is life excluded from ethical consideration or legal protection. Stripped of social protections, Agamben explains, bare life exists as "a living dead man," a mere biological fact left fully exposed to harm. Agamben, *Homo Sacer*, 99.

55. In a similar vein, Aihwa Ong uses the term "graduated" to describe the effects of the "differential state treatment of segments of the population in relation to market calculations, thus intensifying the fragmentation of citizenship." Ong, "Graduated Sovereignty," 57. Differ-

ential treatment grates against middle-class fantasies of being fully incorporated into the public life of the city.

56. Atomic Bunker Romania was a short-lived Romanian-Canadian venture selling private bunkers in Romania. Atomic Bunker Romania, "Atomic Bunker Romania." Although the company closed in 2015 after only a few years in operation, it received regular press coverage in Bucharest and beyond. See, for example, Crişan, "Firmele care construiesc buncăre"; Ochiană, "Altă utilitate"; "Buncăr antiatomic"; "Buncăr subteran." Coverage of Atomic Bunker Romania varied between a curious fascination and biting criticism of the bunker-turned-luxury expenditure. In both instances, reporting played upon the increasingly divergent lives of those who can afford such a private bunker and those who cannot.

57. "Buncăr antiatomic."

58. Koblenko, "Buncăr antiatomic."

59. Ochiană, "Altă utilitate." Masco, *The Future of Fallout*.

60. For an account in the popular press of the flooded civil protection shelters with ventilation systems in disrepair beneath Bucharest's residential blocs, see "Bucureştenii."

61. Dobrescu, "Adăpost antiaerian."

Chapter 10

1. "Un loc de veci."

2. "Bucureşti."

3. Etves, "Locurile de veci."

4. Ivanov, "Piaţa locurilor de veci."

5. O'Neill, "Up, Down, and Away."

6. Rusu, "Privatization of Death," 581.

7. For an encyclopedic account of Bellu Cemetery, covering both those interned and the works of art that commemorate them, see Bezviconi, *Necropola capitalei*. Gheorghe Bezviconi was a prominent historian in Bucharest whose work drew negative attention from the Romanian Communist Party. As a form of persecution, officials stripped Bezviconi of his academic post and assigned him to work as a doorman at Bellu Cemetery. Rather than succumb to boredom, Bezviconi continued in his craft, researching and writing the history of Bucharest's great cemetery. "Povestea lui Bezviconi."

8. Arhivele Naţionale ale României, "Ţara Românească Dosar No. 169." See also Tănase, Manolache, and Filip, *Panteonul naţional* Vol. 2, 10.

9. In the late eighteenth century, noted Michel Foucault, "there arose what could be called an urban fear, a fear of the city, a very characteristic uneasiness: a fear of the workshop and factories being constructed, the crowding together of the population, the excessive height of the buildings, the urban epidemics." Foucault, "Birth of Social Medicine," 144. The putrescence and decay of the city's overfull church cemeteries factored prominently within this growing fear of the city. From Paris (Ariès, *Hour of Our Death*, 57) to London (Joyce, *Rule of Freedom*, 89) to New York (Farland, "Decomposing City," 807), reformers worked to ensure the health and safety of the living by relocating the dead into newly constructed and meticulously organized cemeteries located outside of the city. "Thus, in the outskirts of the cities, at the end of the eighteenth century," Foucault continued, "what appeared was a veritable army of dead people, as perfectly aligned as a regiment being passed in review." Foucault, "Birth of Social Medicine," 147.

10. Tănase, Manolache, and Filip, *Panteonul naţional* Vol. 2, 24.

11. The move from the church's graveyard to the modern cemetery, Philippe Ariès argued, reflects an ideological shift from the ecclesiastical to the secular management of the dead. Ariès, *Hour of Our Death*, 492. Through the modern cemetery, death enters the purview of citizenship by attaching the citizen to the state. McManners, *Death and the Enlightenment*, 359–60.

12. Bezviconi, *Necropola capitalei*, 7–8.

13. As Philippe Ariès astutely observed of Paris's cemeteries, "The topography of the cemetery reproduces the society as a whole, just as a relief map reproduces the contours of a piece of land. All are brought together in the same enclosure, but each has his own place." Ariès, *Hour of Our Death*, 503. In a similar vein, looking across the cemeteries of England, Patrick Joyce adds, "If the cemetery now belonged to everyone, the new cemetery society of this 'everyone' of abstracted individuals reproduced the contradictions inherent in this universalism. The cemetery was riven by class differences from the earliest days." Joyce, *Rule of Freedom*, 91. The same can be said of Bucharest's public cemeteries.

14. "Regarding the Bellu Cemetery as a cultural monument," Gheorghe Bezviconi writes, "we allow ourselves a statement: we do not know of another cemetery in other countries that would singularly concentrate the tombs of its leaders such as has been done at Bellu Cemetery. Historic tombs are scattered in various localities. In the U.S.S.R. there were two capitals; in Germany and Italy more; in France and England there are pantheons, but the great cities have two or more important cemeteries reflecting the longer development of these economic and cultural centers. [In Bellu] we have a cemetery that, more than any other, mirrors the country's past." Bezviconi, *Necropola capitalei*, 38.

15. Aside from the notable figures interned at Bellu, the cemetery contains works by prominent Romanian sculptors and architects, including Karl, Carol, and Frederick Storck, Ion Mincu, Oscar Spathe, Constantin Baraschi, Filip Marin, Raffaello Romanelli, Milița Petrașcu, and Ioan Georgescu. Tănase, Manolache, and Filip, *Panteonul național* Vol. 2: 24, 105–12. For a comprehensive listing of prominent sculptors and their works, see Bezviconi, *Necropola capitalei*, 16–37.

16. "Bellu Cemetery is a true art museum, comprised of the richest collection of sculpture in the country—busts, medallions and statues, made by our old sculptors." Bezviconi, *Necropola capitalei*, 16. The sentiment is shared and expanded upon by others, including Dragomir, *Un București mai puțin cunoscut*, 83; Ilie, *Intre Bellu și Montparnasse*; Tănase, Manolache, and Filip, *Panteonul național* Vol. 2; Oprescu, *Bellu*.

17. As Bezviconi notes, "The very fact that the atheist leaders consecrated by the Academy of the Romanian People's Republic are buried here is a proof of the cultural value of the Bellu Cemetery." Bezviconi, *Necropola capitalei*, 38.

18. As one popular book about Bellu claims: "The rich history of this cemetery, and the numerous personalities that rest here, make Bellu Orthodox a genuine national Pantheon. All over, on the gravestones, crypts or tombs' front piece, one can read the names of scholars, writers, doctors, historians, engineers, artists and in general creators from all fields—as a mark of the palace which eternity reserved for each of them." Tănase, Manolache, and Filip, *Panteonul național* Vol. 2, 11. Similar sentiments are reproduced in Oprescu, *Bellu*, a collection of black and white photographs of Bellu Cemetery produced by Bucharest City Hall. An author is not listed on the title page, nor are page numbers included. The book includes a foreword attributed to the former Mayor of Bucharest, Sorin Oprescu, as well as a brief history of the cemetery and its notable inhabitants. The book presents for tourists and the general public alike the municipality's understanding of Bellu Cemetery and its importance.

19. The cult of the dead in Romanian Orthodoxy entails a series of post-funeral rituals, which last for at least seven years. These commemorative ceremonies are performed in a religious framework and blessed by the priest. They can be classified according to three types, the simplest of which are the *pomenire* (remembering), which occur during normal liturgical services. The second is called *parastas*, which entails a short religious service as well as the eating of a ritual sweet bread that is shared among participating family and friends. The ceremony occurs either inside the church or at the grave. It takes place three, nine, twenty-one, and forty days after the funeral; then three, six, and nine months after; and then annually for seven years from the date of the death. The third ceremony is called *pomană*, which occurs after the burial, the following day, and then after six weeks, six months, and every year until the seventh. About forty to one hundred people participate in a typical *pomană*. Geană, "Remembering Ancestors," 351–52. In Romanian folklore, the rituals associated with death are organized around Romanian Orthodoxy but also contain pre-Christian beliefs and practices. See Kligman, *Wedding of the Dead*, Chapter 3.

20. Though the rituals themselves are overseen by priests, Maria Bucur makes clear, women are overwhelmingly responsible for the proper organization of mourning rituals, both customarily and normatively. Bucur, "Gender and Religiosity," 31.

21. Kligman, *Wedding of the Dead*, 214. Similarly, Katherine Verdery argues that corpses are not meaningful in themselves but rather become meaningful through the cosmologies and practices that relate the living and the dead. Verdery, *Political Lives of Dead Bodies*, 26.

22. Since concluding this research, the municipality has extended the Metro's M4 line so that it now reaches Străulești, reducing travel time from the center to about thirty minutes.

23. The modern cemetery introduced new dispositional techniques that made social differentiation possible in death. Reformers organized the modern cemetery through the enclosure of its borders, the partitioning of its plots, the creation of functional emplacements to ensure public health, and the introduction of the art of "rank," so that each corpse received a precise relational position. Johnson, "Modern Cemetery," 781. As Foucault argued, this period of modernization brought the individualized cemetery into existence, with the normalization of the individualized coffin and the family tomb. Foucault, "Birth of Social Medicine," 147.

24. "Cimitirul Bellu."

25. Oprescu, *Bellu*.

26. "Cimitirul Bellu."

27. "Cimitirul Bellu."

28. Răduță, "Spectacol printre morminte"; Pârvulescu, "Noaptea muzeelor."

29. Samoila, "Morții dintre vii."

30. Foucault, "Of Other Spaces," 25.

31. In this way, the politics of life and of death merge. See Mbembe, *Necropolitics*.

32. "To the socially recognized hierarchy of the arts, and within each of them, of genres, schools, or periods, corresponds a social hierarchy of the consumers. This predisposes tastes to function as markers of 'class,'" wrote Pierre Bourdieu, *Distinction*, 1–2. One articulation of taste that contrasts with the monumentalism of Romanian cemeteries is the "aesthetic restraint" of the modern American cemetery. See Dawdy, "Archaeology of Modern American Death," 456.

33. Rusu, "Privatization of Death," 576.

34. Lica, "Metrul pătrat de țărână."

35. "Eternitate dată în rate"; "București."

36. Take Gail Kligman's meticulous account of burial practices in Romania from the 1980s: "Most funeral expenses are incurred during this preparatory phase. The priest (or priests) will be paid a certain fee, as will the cook, bell-ringer, and coffin-maker. Food, drink, and candles must be purchased; the coffin and the flower wreaths must be bought; flags and rugs must be rented from the church. Additional expenses include the costs of the ritual meals for the obligatory mourning days, as well as of a gravestone or carved cross. Like a wedding, a funeral is expensive (for example, a large funeral cost one family 5,900 lei, over two months' salary)." Kligman, *Wedding of the Dead*, 173. At that moment in history, matters of real estate did not factor into the accounting in any mentionable way.

37. As the likes of Friedrich Hayek warn, "Our freedom of choice in a competitive society rests on the fact that, if one person refuses to satisfy our wishes we can turn to another. But if we face a monopolist we are at his mercy." Hayek, *Road to Serfdom*, 96. A state monopoly concerned Hayek most of all. As Michel Foucault argued, the neoliberal freedom of choice pursued within a market of options functions as a critique of public authorities, working to ensure that the state does not govern too much. Foucault, *Birth of Biopolitics*, 246.

38. Personal correspondence, January 25, 2016.

39. Nicolae Grigorescu was a Romanian painter trained in Paris. He is credited with introducing impressionism to Romania and is considered one of the founders of modern Romanian painting. See Varga, *Nicolae Grigorescu*.

40. "Bucureşti."

41. "Mafia mormintelor."

Conclusion

1. For North American readers, Lucian's use of "flee" no doubt evokes images of white flight, in which upwardly mobile whites fled decaying U.S. cities for the suburbs in the mid- to late twentieth century. See Jackson, *Crabgrass Frontier*. The process reflected a brutal racial and class politics that excluded Black Americans. See Johnson, *Broken Heart of America*. Importantly, in the context of Bucharest, the middle classes are not trying to escape a deindustrialized city with a collapsing economy. To the contrary, they are seeking to remain connected to a postindustrial city whose economy is expanding. Their "flight" is from the periphery to the center rather than the reverse.

2. Popescu, "Norii ruginii."

3. For the public, see Habermas, *Structural Transformation*, as discussed in Chapter 8. As Benedict Anderson famously argued, "regardless of the actual inequality and exploitation that may prevail . . . the nation is always conceived as a deep, horizontal comradeship. Ultimately it is this fraternity that makes it possible, over the past two centuries, for so many millions of people, not so much to kill, as willingly to die for such limited imaginaries." Anderson, *Imagined Communities*, 7. In a similar vein, Victor Turner coined the term *communitas* to capture an essential model of human interrelatedness—a sense "of society as an unstructured or rudimentarily structured and relatively undifferentiated *comitatus*, community, or even communion of equal individuals . . . without which there could be no society." Turner, "Liminality and Communitas," 96. Finally, Émile Durkheim developed "collective effervescence" to capture the affective bonds that build community. Durkheim, *Elementary Forms*, 213

4. The question of what counts as "here," Jim Ferguson observes, is undertheorized. As Ferguson writes, "In recent decades, we have spent a lot of critical and political energy rethinking who counts as 'us.' We have recognized that taken-for-granted social identities are in fact

elaborate and consequential constructions, whose reworking must be at the heart of our politics. We now need an equal dedication to the problem of what counts as 'here'—recognizing
that commonsense notions of presence (the idea that some people are 'here, among us' while
others are not) are also elaborate constructions." Ferguson, *Presence and Social Obligation*,
42. The intent of this book is to add depth to the concept of "here" by bringing into view the
city's ranked, vertical ordering. The underground may create space for more people to be
"here," in the city, but those present on bottom do not belong "here" in the same way as those
present on top of the city.

5. The horizontal tension between center and periphery was popularized by the Chicago
School. See Park and Burgess, *The City*. Later, a decentered anthropology shifted attention away
from the center and the periphery to account for the rise of exclusive, gated communities across
the city. Caldeira, *City of Walls*; Low, *Behind the Gates*.

6. Kim, "Extreme Primacy of Location."

7. See Endicott, Johnston, and Lin, *Underground Cities*; VisitSeoul.net, "Seoul Underground"; Tousignant, "Guide to the Underground City."

8. Dymond, "Creating an Underground Library."

BIBLIOGRAPHY

"#undergroundbar." *www.instagram.com*. 2023. https://www.instagram.com/explore/tags /undergroundbar/.

5 to Go. "Franchise." *www.fivetogo.ro*. 2021. https://fivetogo.ro/en/franciza/#.

"3.000 de maşini abandonate ocupă locurile de parcare ale bucureştenilor." *Digi24.Ro*. January 9, 2018.

"1,3 miliarde litri/lună—Pierderile de apă din reţeaua RADET în sezonul de iarnă." *Media-Fax.Ro*. 2019. https://www.mediafax.ro/social/1-3-miliarde-litri-luna-pierderile-de-apa -din-reteaua-radet-in-sezonul-de-iarna-18405824.

ABC News. "'The Shame of a Nation,' Interview by Barbara Walters and Hugh Downs." *20/20*. 1989.

"About SkyTower." *Skytower.ro*. 2018. http://www.skytower.ro/about-skytower/.

Abufarha, Nasser. *The Making of a Human Bomb: An Ethnography of Palestinian Resistance*. Durham, NC: Duke University Press, 2009.

ACUUS. "Cooperation Agreement between UN-HABITAT and ACUUS." *acuus.org*. 2021.

"Adio covrigilor şi gogoşilor de la metrou. Zici că eşti la Istanbul!" *Wall-Street.Ro*. 2011. https:// www.wall-street.ro/articol/Actualitate/104562/Adio-covrigilor-si-gogosilor-de-la -metrou-Zici-ca-esti-la-Istanbul.html#gref.

Adler, Jerry. "The Next Frontier in Urban Design Will Send You Underground." *Smithsonian Magazine*. Washington, D.C., December 2013. https://www.smithsonianmag.com/arts -culture/the-next-frontier-in-urban-design-will-send-you-undeground-180947628/.

Admiraal, Han and Shipra Narang Suri. *Think Deep: Planning, Development and Use of Underground Space in Cities*. Netherlands: ISOCARP, 2015.

ADRBI. "Planul de dezvoltare regională 2014–2020." Bucharest, 2015.

Agamben, Giorgio. *Homo Sacer: Sovereign Power and Bare Life*. Stanford, CA: Stanford University Press, 1998.

Alexandrescu, Horia. "'Oraşul' de sub oraş." *Tineretul Liber*. 1990.

Alexe, Anca. "CBRE: Bucharest Office Space Market Needs New Projects as Demand Keeps Growing." *Business Review*. May 31, 2018.

Alexe, Iris et al. "Social Impact of Emigration and Rural-Urban Migration in Central and Eastern Europe: Executive Summary." Romania: European Commission, 2012.

Alexe, Roxana. "Parcarea subterană de la Universitate." *MediaFax*. 2012. http://www.mediafax .ro/social/parcarea-subterana-de-la-universitate-inaugurata-cum-arata-si-care-sunt -preturile-pentru-acces-galerie-foto-10324468.

"Amenajarea Pieţei Universităţii, finalizată în iunie 2012. Vezi proiectele de pe podium." *B365. Ro*. 2011. http://www.b365.ro/amenajarea-pietei-universitatii-finalizata-in-iunie-2012-vezi -cum-va-arata_116522.html.

Anderson, Benedict R. *Imagined Communities: Reflections on the Origin and Spread of Nationalism*. New York: Verso, 2006.

Appadurai, Arjun. *Modernity at Large: Cultural Dimensions of Globalization*. Public Worlds, Vol. 1. Minneapolis: University of Minnesota Press, 1996.

Argenbright, Robert. "Avtomobilschchina: Driven to the Brink in Moscow." *Urban Geography* 29, no. 7 (2008): 683–704.

Arhivele Naționale ale României. "Serviciul arhitecturii și oficiul tehnic: 21/1944." Bucharest, 1944.

———. "Țara Românească Dosar No. 169." Bucharest, 1865.

Ariès, Philippe. *The Hour of Our Death*. Oxford: Oxford University Press, 1981.

"Așa arată un buncăr antiatomic de ultimă generație." *www.Tion.Ro*. 2014. https://www.tion.ro /stirile-judetului-timis/asa-arata-un-buncar-anti-atomic-de-ultima-generatie-193063/.

"Atenție, se ridică mașinile abandonate din sectorul 1! Cât costă să fie recuperate." *Ziare.Com*. 2017.

Atkinson, Rowland. "Necrotecture: Lifeless Dwellings and London's Super-Rich." *International Journal of Urban and Regional Research* 43, no. 1 (2019): 2–13.

Atomic Bunker Romania. "Atomic Bunker Romania." *http://atomicbunker.ro*. 2018. http:// atomicbunker.ro/.

"Atribuții extinse pentru polițiștii locali: Să dirijeze traficul și să dea amenzi—Proiect." *Digi 24.Ro*. 2017. http://s2.ziareromania.ro/?mmid=dd05ce911ca0f1876.

Augé, Marc. *In the Metro*. Minneapolis: University of Minnesota Press, 2002.

Augustine of Hippo. *Confessions: Books I-XIII*. Indianapolis, IN: Hackett Publishing Company, 1993.

"Average Net Income Reaches EUR 657 in Bucharest." *Romania-Insider.Com*. March 22, 2016. https://www.romania-insider.com/average-net-income-reaches-eur-657-in-bucharest/.

Bachelard, Gaston. *The Poetics of Space*. New York: Beacon Press, 1994.

Baker, Mark. "Bucharest's New Old City." *BBC Travel*. July 16, 2013. http://www.bbc.com/travel /story/20130712-bucharests-new-old-city.

Bakhtin, Mikhail M. *Rabelais and His World*. Bloomington, IN: Indiana University Press, 1984.

Băltărețu, Răzvan. "Biblioteca Digitală revine cu peste 500 de titluri la metrou cu ajutorul Vodafone România și al editurii Humanitas." *Adevărul.Ro*. 2013. https://adevarul.ro/tech /internet/biblioteca-digitala-revine-500-titluri-metrou-ajutorul-vodafone-romania -editurii-humanitas-1_511b55fe344a78211835492f/index.html.

Bănilă, Silviu. "Operatorul metroului din Beijing s-a inspirat din stația Piața Victoriei din București și le-a făcut călătorilor chinezi o surpriză." *Gândul.Info*. 2015. https://www .gandul.info/magazin/operatorul-metroului-din-beijing-s-a-inspirat-din-statia-piata -victoriei-din-bucuresti-si-le-a-facut-calatorilor-chinezi-o-surpriza-14015871.

Banister, C. E. "Transport in Romania—a British Perspective." *Transport Reviews* 1, no. 3 (1981): 251–70.

Barbu, Iulia. "Lacul a scăzut din cauza unui bloc-turn." *Antena3.Ro*. 2009. https://jurnalul .antena3.ro/stire-special/lacul-a-scazut-din-cauza-unui-bloc-turn-503026.html.

Baudrillard, Jean. "Modernity." *CTheory* 11, no. 3 (1987): 63–72.

BDO. "Cyberfort Group." *Bdo.Co.Uk*. September 2018. https://www.bdo.co.uk/en-gb/plugdin /interviews/cyberfort-group.

Bear, Laura. "Anthropological Futures: For a Critical Political Economy of Capitalist Time. ASA's Raymond Firth Lecture, 2016." *Social Anthropology* 25, no. 2 (2017): 142–58.

———. "Time as Technique." *Annual Review of Anthropology* 45 (2016): 487–502. https://doi
.org/10.1146/annurev-anthro-102313-030159.

Bell, Laird. "Air Rights." *Illinois Law Review* 23, no. 3 (1928): 250–64. www.journal.uta45jakarta
.ac.id.

Belzberg, Edet. (Director). (2001). *Children Underground.* [Documentary]. http://www.imdb
.com/title/tt0264476/?ref_=fn_al_tt_1.

Ben-Joseph, Eran. *ReThinking a Lot: The Design and Culture of Parking.* Cambridge, MA: MIT
Press, 2012.

Benjamin, Walter. *The Arcades Project.* Cambridge, MA: Belknap Press of Harvard University Press, 2002.

Bennett, Luke. "Grubbing out the Führerbunker: Ruination, Demolition, and Berlin's Difficult Subterranean Heritage." *Geographia Polonica* 92, no. 1 (2019): 71–82.

———. "The Bunker's After-Life: Cultural Production in the Ruins of the Cold War." *Journal
of War and Culture Studies* 13, no. 1 (2020): 1–10.

Berger Ziauddin, Silvia. "(De)Territorializing the Home. The Nuclear Bomb Shelter as a
Malleable Site of Passage." *Environment and Planning D: Society and Space* 35, no. 4
(2017): 674–93.

Berlant, Lauren. *Cruel Optimism.* Durham, NC: Duke University Press, 2011.

Bertacchini, Enrico E. and Donatella Saccone. "Toward a Political Economy of World Heritage." *Journal of Cultural Economics* 36, no. 4 (2012): 327–52.

Beznilă, Gheorghe. "Stația Piața Unirii." *Arhitectura* 4 (1980): 19.

Bezviconi, Gheorghe. *Necropola capitalei: dicționar enciclopedic.* Bucharest: Institutul de Is-
torie "N. Iorga," 1972.

Biehl, João Guilherme. *Vita: Life in a Zone of Social Abandonment.* Berkeley: University of California Press, 2005.

Billé, Franck. *Voluminous States: Sovereignty, Materiality, and the Territorial Imagination.* Durham, NC: Duke University Press, 2020.

Bitetti, B. et al. "Construction Methods, Monitoring and Follow-up of TBM Tunnelling beneath an Operating Metro Station in Bucharest." In *Tunnels and Underground Cities:
Engineering and Innovation Meet Archaeology, Architecture and Art,* edited by Daniele
Peila, Giulia Viggiani, Tarcisio Celestino, 5369–79. London: Taylor & Francis Group,
2019.

Blănaru, Cristina. "McCann Erickson Bucharest—Silver in Promo and Bronze in Direct; First
Winners of Cannes 2013 Announced." *(Ad)Hugger.Net.* June 18, 2013. http://www.adhugger
.net/2013/06/18/mccann-erickson-bucharest-silver-in-promo-and-bronze-in-direct-first
-winners-of-cannes-2013-announced/.

Bloodworth, Sandra and William S. Ayres. *New York's Underground Art Museum: MTA Arts
& Design.* New York: Monacelli Press, 2014.

Boas, Franz. "The Study of Geography." *Science* 9, no. 210 (1887): 137–41.

Böhme, Gernot. *Critique of Aesthetic Capitalism.* Berlin: Mimesis International, 2017.

———. "The Art of the Stage Set as a Paradigm for an Aesthetics of Atmospheres." *Ambiances*
(February 2013): 2–8.

———. "The Atmosphere of a City." *Issues in Contemporary Culture and Aesthetics* 7 (1998):
5–13.

Boia, Lucian. *History and Myth in Romanian Consciousness.* Hors Collection. Budapest:
Central European University Press, 2001.

Bonini Baraldi, Sara, Daniel David Shoup, and Luca Zan. "When Megaprojects Meet Archaeology: A Research Framework and Case Study from Yenikapi, Istanbul." *International
Journal of Cultural Policy* 25, no. 4 (June 7, 2019): 423–44.

Bourdieu, Pierre. *Distinction: A Social Critique of the Judgment of Taste.* Cambridge, MA: Harvard University Press, 1987.

———. "Site Effects." In *The Weight of the World: Social Suffering in Contemporary Society*,
edited by Pierre Bourdieu, 123–30. Stanford, CA: Stanford University Press, 1999.

Bourgois, Philippe. *In Search of Respect: Selling Crack in El Barrio.* 2nd ed. Cambridge: Cambridge University Press, 2003.

Bowers, Rachael and Kevin Booth. "Preserving and Managing York Cold War Bunker: Authenticity, Curation and the Visitor Experience." In *In the Ruins of the Cold War Bunker
Affect, Materiality and Meaning Making*, edited by Luke Bennett, 201–14. New York: Row-
man & Littlefield International, 2017.

Boyarin, Jonathan. "Placing Reading: Ancient Israel and Medieval Europe." In *The Ethnography of Reading*, edited by Jonathan Boyarin. Berkeley: University of California Press,
1993.

———. *The Ethnography of Reading.* Berkeley: University of California Press, 1993.

Bridger, Jessica. "Empty Libraries, Captive Subway Audiences." *Uncube.* March 14, 2013. http://
www.uncubemagazine.com/blog/8757145.

Brzostek, Błażej. "Romania's Peculiar Way in the Landscapes of Bucharest, 1806–1906." *Acta
Poloniae Historica* 111 (2015).

Buck-Morss, Susan. *The Dialectics of Seeing: Walter Benjamin and the Arcades Project.* Cambridge, MA: MIT Press, 1991.

———. "The Flaneur, the Sandwichman and the Whore: The Politics of Loitering." *New German Critique* no. 39 (1986): 99–140.

Bucknell, Alice. "This Nazi-Era Bunker Is Now a Thriving Art Gallery." *Architecturaldigest
.Com.* January 2018. https://www.architecturaldigest.com/story/nazi-era-bunker-now
-thriving-art-gallery.

Bucur, Alexandru. "Acte normativ-legislative privind protecția civilă din România (1916–
1945)." *Cultura Mediesană* II (2013): 59–63.

Bucur, Maria. "Gender and Religiosity among the Orthodox Christians in Romania: Conti-
nuity and Change, 1945–1989." *Aspasia* 5, no. 1 (2011): 28–45.

"București—Orașul unde un loc de veci costă cât o casă. 'Moartea este un lux.'" *Antena3.Ro.*
2012. https://www.antena3.ro/actualitate/bucuresti-orasul-unde-un-loc-de-veci-costa-cat
-o-casa-moartea-este-un-lux-168282.html.

"Bucureștenii nu au unde se adăposti în caz de calamitate." *Digi24.Ro.* 2015. https://www.digi24
.ro/stiri/actualitate/social/bucurestenii-nu-au-unde-se-adaposti-in-caz-de-calamitate
-442230.

Bunker-51.com. "Bunker 51." *bunker-51.com.* 2020.

The Bunker. "The Bunker." *https://www.thebunker.net.* 2020.

Burawoy, Michael. "Facing an Unequal World." *Current Sociology* 63, no. 1 (2015): 5–34.

Burcuș, S. T. "O vizită la un amator de artă și arhitectură Rromă." *Arhitectura* 1–2 (1906):
29–32.

Burke, Patrick. *Come in and Hear the Truth: Jazz and Race on 52nd Street.* Chicago: University of Chicago Press, 2008.

Bürkner, Hans Joachim and Silviu G. Totelecan. "Assemblages of Urban Leisure Culture in Inner-City Bucharest." *International Journal of Urban and Regional Research* 42, no. 5 (2018): 786–806.

Burrows, Roger, Stephen Graham, and Alexander Wilson. "Bunkering down? The Geography of Elite Residential Basement Development in London." *Urban Geography* 43, no. 9 (2022): 1372–1393.

Caldeira, Teresa P. R. *City of Walls: Crime, Segregation, and Citizenship in São Paulo.* Berkeley: University of California Press, 2001.

Caragiale, Mateiu I. *Gallants of the Old Court: A Novel*, translated by Christian Baciu. Bucharest, Romania: eLiteratura, 2013.

Caraher, William. "Slow Archaeology: Technology, Efficiency, and Archaeological Work." In *Mobilizing the Past for a Digital Future: The Potential of Digital Archaeology*, edited by Derek B. Counts, Erin Walcek Averett, and Jody Michael Gordon, 421–42. Open Access E-Books. Grand Forks, ND: Digital Press at the University of North Dakota, 2016.

Cărtărescu, Mircea. *Nostalgia: A Novel.* New York: New Directions Book, 2005.

Castells, Manuel. *The Rise of the Network Society: The Information Age: Economy, Society, and Culture.* Information Age Series. London: Wiley, 2011.

"Ceaușescu's Nuclear Bunker: The Last Secret of Romania's Communist Dictator." *Agencia EFE.* March 23, 2017. https://www.efe.com/efe/english/eventos/ceausescu-s-nuclear -bunker-the-last-secret-of-romania-communist-dictator/50000272-3216881.

Céline, Louis-Ferdinand. *Journey to the End of the Night.* New York: New Directions Book, 2006.

Certeau, Michel de. *The Practice of Everyday Life.* Berkeley: University of California Press, 2002.

Chayka, Kyle. *The Longing for Less: Living with Minimalism.* New York: Bloomsbury Publishing, 2020.

Chelcea, Liviu. "Ancestors, Domestic Groups, and the Socialist State: Housing Nationalization and Restitution in Romania." *Comparative Studies in Society and History* 45, no. 04 (November 2003): 714–40.

———. "Marginal Groups in Central Places: Gentrification, Property Rights, and Post-Socialist Primitive Accumulation." In *Social Changes and Social Sustainability in Historical Urban Centers: The Case of Central Europe*, edited by Gyorgy Enyrdi and Zoltan Kovacs, 127–46. Hungary: Centre for Regional Studies of the Hungarian Academy of Sciences, 2006.

———. "Post-Socialist Acceleration: Fantasy Time in a Multinational Bank." *Time & Society* 24, no. 3 (December 22, 2015): 348–66.

Chelcea, Liviu and Oana Druță. "Zombie Socialism and the Rise of Neoliberalism in Post-Socialist Central and Eastern Europe." *Eurasian Geography and Economics* 57, no. 4–5 (2016): 521–44.

Chelcea, Liviu and Ioana Iancu. "An Anthropology of Parking: Work, Infrastructures of Automobility and Circulation." *Anthropology of Work Review* 36, no. 2 (2015): 62–73.

Chelcea, Liviu, Raluca Popescu, and Darie Cristea. "Who Are the Gentrifiers and How Do They Change Central City Neighbourhoods? Privatization, Commodification, and Gentrification in Bucharest." *Geografie* 120, no. 2 (2015): 113–33.

Chen, Stefanos. "New Supertalls Test the Limits, as the City Consults an Aging Playbook." *New York Times.* October 1, 2021. https://www.nytimes.com/2021/10/01/realestate/supertalls -safety.html.

——. "The Underground Apartment Market." *New York Times*. August 17, 2018. https://
    www.nytimes.com/2018/08/17/realestate/the-underground-apartment-market.html.
Chrisafis, Angelique. "Paris Hopes €1bn Revamp of Les Halles Can Become City's 'Beating
    Heart.'" *Guardian*. February 1, 2016. https://www.theguardian.com/world/2016/feb/01
    /paris-hopes-1bn-revamp-of-les-halles-can-become-citys-beating-heart.
Churchward, Phil, Brian Klein, and Owen Trevor. "Top Gear: Episode 14.1 (Romania)." U.K.,
    2010.
Cicovschi, Afrodita. "Metrorex începe evacuarea spațiilor comerciale de la metrou. Contractul
    încheiat cu firma liderului de sindicat a expirat din 2018." *Adevărul.Ro.* 2021. https://adevarul
    .ro/economie/stiri-economice/metrorex-incepe-evacuarea-spatiilor-comerciale-metrou
    -contractul-incheiat-firma-liderului-sindicat-expirat-2018-1_606aaa025163ec4271a0b2ab
    /index.html.
"Cimitirul Bellu, deschis sâmbătă noaptea pentru turiști." *Stirileprotv.Ro.* 2011. https://
    stirileprotv.ro/stiri/social/cimitirul-bellu-deschis-sambata-noaptea-pentru-turisti
    -galerie-foto.html.
"Cimitirul Bellu va fi inclus pe harta obiectivelor cultural-turistice Europene." *Adevărul.
    Ro.* 2010. https://adevarul.ro/news/eveniment/cimitirul-bellu-inclus-harta-obiectivelor
    -cultural-turistice-europene-1_50aecbc97c42d5a663a08ad0/index.html.
"Cimitirul Bellu va fi transformat în muzeu în aer liber." *Metropotam.* 2014. http://metropotam
    .ro/La-zi/cimitirul-bellu-va-fi-transformat-in-muzeu-in-aer-liber-art3534651501/.
Ciortan, S T. "Casa Octav Gușerescu." *Arhitectura* 2 (1916): 65.
City of New York Press Office. June 28, 2018. "De Blasio Administration Launches Pilot Base-
    ment Conversion Program in East New York; Legislation Proposed to Modify Building
    Code Standards for Area." [Press Release] https://www1.nyc.gov, 2018. https://www1.nyc
    .gov/office-of-the-mayor/news/337-18/de-blasio-administration-launches-pilot
    -basement-conversion-program-east-new-york-legislation.
"La Claustra." http://www.claustra.ch, 2021. http://www.claustra.ch.
Colliers International. "Research & Forecast Report." Bucharest, 2018.
——. "Romania: Research and Forecast Report for Mid-Year 2012." Bucharest, 2012.
Comaroff, Joshua. "On the Materialities of Air." *City* 21, no. 5 (September 3, 2017): 607–13.
Rocher, Stijn and Michael H. Stierle. "Household Saving Rates in the EU: Why Do They Differ
    So Much?" Vol. 5. Luxembourg: European Commission, 2015.
Consiliul General al Municipiului București. "Master planul general pentru transport urban—
    București." Bucharest, Romania, 2008.
Contreras, Randol. *The Stickup Kids: Race, Drugs, Violence, and the American Dream.* Berke-
    ley: University of California Press, 2012.
Council of Europe. *European Convention on the Protection of the Archaeological Heritage (Re-
    vised).*1992.   https://rm.coe.int/CoERMPublicCommonSearchServices/DisplayDCTM
    Content?documentId=090000168007bd25.
Council on Tall Buildings and Urban Habitat. "Euro Tower: Bucharest, Romania." *https://www
    .skyscrapercenter.com.* 2021. https://www.skyscrapercenter.com/building/euro-tower/11304.
——. "Petronas Twin Tower 1." *https://www.skyscrapercenter.com.* 2021. https://www
    .skyscrapercenter.com/building/petronas-twin-tower-1/149.
Cozmei, Victor. "Ce s-a întâmplat la metrou: de ce s-a surpat pământul la Eroilor, acolo unde
    sapă 'cârtițele' de pe magistrala 5." *HotNews.Ro.* 2015. https://monitorizari.hotnews.ro/stiri

-infrastructura_articole-20665433-intamplat-metrou-surpat-pamantul-eroilor-acolo
-unde-sapa-cartitele-magistrala-5.htm.

———. "Ce probleme reclamă șoferii la parcarea subterană de la Universitate." *HotNews.Ro.* 2013.

Craciun, Oana. "Elevii din 300 de licee din țară vor putea "răsfoi" online 100 de cărți din biblioteca digitală gratuită creată de Vodafone." *Adevărul.Ro.* 2013. https://adevarul.ro /educatie/scoala/vodafone-1_5267d4cfc7b855ff566fa2e7/index.html.

Crang, Mike. "Rhythms of the City: Temporalised Space and Motion." In *Timespace: Geographies of Temporality,* edited by Jon May and Nigel Thrift, 187–207. London: Routledge, 2003.

Creangă-Stoilești, Gheorghe F. *Istoria apărării civile.* Galați, România: Porto-Franco, 1993.

Crișan, Andrea. "Firmele care construiesc buncăre primesc mai multe cereri." *Digi24.Ro.* 2014. https://www.digi24.ro/regional/digi24-oradea/firmele-care-construiesc-buncare -primesc-mai-multe-cereri-245827.

Cristescu, George-Andrei. "Blocaj la metrou: Aglomerație pe ruta Berceni—Pipera, la oră de vârf." *Adevărul.Ro.* 2019. https://adevarul.ro/news/bucuresti/video-blocaj-metrou-aglo meratie-ruta-berceni-pipera-ora-varf-1_5ce4dd06445219c57ee7dcac/index.html.

Crowley, David and Susan E Reid. *Socialist Spaces: Sites of Everyday Life in the Eastern Bloc.* London: Berg, 2002.

"Cum arată după reabilitare parcarea de la Teatrul Național. Investiție de 3 MIL. Euro." *Wall-Street.Ro.* 2015. https://www.wall-street.ro/articol/Auto/180411/cum-arata-dupa-reabilitare -parcarea-de-la-teatrul-national-investitie-de-3-mil-euro.html#gref.

"Cum arată rețeaua de buncăre antiatomice de sub București." *Romaniatv.Net.* 2016. https:// www.romaniatv.net/cum-arata-reteaua-de-buncare-antiatomice-de-sub-bucuresti-ce -se-afla-in-buncarul-demnitarilor_266568.html.

Curea, Mirel. "Planul Z a funcționat până la deciderea judecării cuplului." *Evenimentul Zilei.* 1993.

Daly, Ian. "Nuclear Bunker Houses World's Toughest Server Farm." *Wired.Co.Uk.* October 2010. https://www.wired.co.uk/article/20-thousand-terabytes-under-the-ground.

Danciu, Aniela Raluca and Mihaela Gruiescu. "The Evolution of FDI in Romania During the Period 1990–2009." *Romanian Economic and Business Review* 6, no. 1 (2013): 97–103.

Danta, Darrick. "Ceaușescu's Bucharest." *Geographical Review* 83, no. 2 (April 1, 1993): 170–82.

Davies, Robert. "PULMAN: Public Libraries, Local Museums and Archives Learning from Each Other in EEurope." *Alexandria* 15, no. 3 (2003): 151–69.

Dawdy, Shannon Lee. "Archaeology of Modern American Death: Grave Goods and Blithe Mementoes." In *The Oxford Handbook of the Archaeology of the Contemporary World,* edited by Paul Graves-Brown, Rodney Harrison, and Angela Piccini, 451–65. Oxford: Oxford University Press, 2013.

"De ce parcarea Străulești, o investiție de sute de milioane de lei, este pustie." *Digi24.Ro.* n.d. https://www.digi24.ro/stiri/economie/transporturi/parcarea-straulesti-inutila -investitia-de-sute-de-milioane-de-lei-este-pustie-1066857.

Deletant, Dennis. *Ceaușescu and the Securitate: Coercion and Dissent in Romania, 1965–1989.* London: M. E. Sharpe, 1995.

Deleuze, Gilles. *Foucault.* Minneapolis: University of Minnesota Press, 1988.

Deleuze, Gilles and Félix Guattari. *A Thousand Plateaus: Capitalism and Schizophrenia.* Minneapolis: University of Minnesota Press, 1987.

Delmastro, Chiara, Evasio Lavagno, and Laura Schranz. "Underground Urbanism: Master Plans and Sectorial Plans." *Tunnelling and Underground Space Technology* 55 (2016): 103–11.

Demekas, Dimitri and Mohsin S. Khan. "The Romanian Economic Reform Program." Occasional Paper—International Monetary Fund. Washington, D.C.: International Monetary Fund, 1991

Deoancă, Adrian. "Class Politics and Romania's 'White Revolution.'" *Anthropology News* 58, no. 4 (2017): e394–98.

Departamentul pentru Situații de Urgență (DSU). "Hartă: Adăposturi de protecție civilă." DSU App for iPhone, 2021.

"Descoperirile arheologice schimbă proiectul parcării de la Universitate." *Adevărul.Ro.* 2010. https://adevarul.ro/news/bucuresti/descoperirile-arheologice-schimba-proiectul -parcarii-universitate-1_50bdec2a7c42d5a663d00ad2/index.html.

Dobrescu, Petre. "Știai că sub blocul tău ar putea fi un adăpost antiaerian?" *Libertatea.Ro.* 2014. https://www.libertatea.ro/stiri/exclusiv-stiai-ca-sub-blocul-tau-ar-putea-fi-un-adapost -antiaerian-1067555.

———. "Update: Primul termen în dosarul Colectiv a început cu o amânare." *Libertatea.Ro.* 2017. https://www.libertatea.ro/stiri/incepe-judecata-dosarul-colectiv-1735959.

Doroftei, Marius. "Incendiu în Colectiv, un pompier explică de ce flăcările au cuprins clubul atât de rapid: Era mai bine în iad decât acolo." *Ziare.Com.* 2015. https://ziare.com/stiri /incendiu-club-colectiv/incendiu-in-colectiv-un-pompier-explica-de-ce-flacarile-au -cuprins-clubul-atat-de-rapid-era-mai-bine-in-iad-decat-acolo-1391081.

Dospinescu, Andrei and Giuseppe Russo. "Romania: Systematic Country Diagnostic— Background Note, Migration." Washington, D.C., 2018. https://documents1.worldbank .org/curated/en/210481530907970911/pdf/128064-SCD-PUBLIC-P160439-RomaniaSCD BackgroundNoteMigration.pdf.

Douglas, Mary. *Purity and Danger: An Analysis of Concepts of Pollution and Taboo.* London: Taylor, 2002.

Dragomir, Silviu. *Un București mai puțin cunoscut.* Bucharest: Lucman, 2008.

Dreger, Christian et al. "Wage and Income Inequality in the European Union." Brussels: European Parliament, 2015. http://www.europarl.europa.eu/RegData/etudes/STUD/2015 /536294/IPOL_STU(2015)536294_EN.pdf.

Dresch, P. "Segmentation: Its Roots in Arabia and its Flowering Elsewhere." *Cultural Anthropology* 3, no. 1 (1988): 50–67.

Dumitrache, Liliana, Daniela Zamfir, and Mariana Nae. "The Urban Nexus: Contradictions and Dilemmas of (Post)Communist (Sub)Urbanization in Romania." *Human Geographies-Journal of Studies and Research in Human Geography* 10, no. 1 (2016): 39–58.

Dumitrescu, Andrei. "Ancheta incendiului din Colectiv—PICCJ: Cauzele incendiului—Show pirotehnic în spațiu impropriu, număr mare de persoane. Patronii clubului au fost reținuți. Ei sunt acuzați de ucidere din culpă." *MediaFax.Ro.* 2015. https://www.mediafax.ro/social /ancheta-incendiului-din-colectiv-piccj-cauzele-incendiului-show-pirotehnic-in-spatiu -impropriu-numar-mare-de-persoane-patronii-clubului-au-fost-retinuti-ei-sunt -acuzati-de-ucidere-din-culpa-foto-14870584.

Dumont, Louis. "Preface by Louis Dumont to the French Edition of The Nuer." In *Studies in Social Anthropology: Essays in Memory of E.E. Evans-Pritchard*, edited by John Beattie and R. Godfrey Lienhardt, 328–42. Oxford: Clarendon Press, 1975.

Dunn, Elizabeth C. *Privatizing Poland: Baby Food, Big Business, and the Remaking of Labor.* Ithaca, NY: Cornell University Press, 2004.

———. "Standards and Person-Making in East Central Europe." In *Global Assemblages: Technology, Politics, and Ethics as Anthropological Problems*, edited by Aihwa Ong and Stephen J. Collier, 173–193. London: Wiley, 2004.

Durkheim, Émile. *The Elementary Forms of Religious Life.* New York: Oxford University Press, 2001.

———. *The Rules of Sociological Method.* New York: Free Press, 1982.

Dymond, Cameron. "Creating an Underground Library in an Inner City Marshland." *ARUP .org.* 2020. https://www.arup.com/projects/green-square-library.

Elborough, T. and A. Horsfield. *Atlas of Improbable Places: A Journey to the World's Most Unusual Corners.* London: Aurum Press, 2016.

Elden, Stuart. "Secure the Volume: Vertical Geopolitics and the Depth of Power." *Political Geography* 34 (2013): 35–51.

Eliade, Mircea. *The Old Man and the Bureaucrats.* South Bend, IN: University of Notre Dame Press, 1979.

———. *The Sacred and the Profane: The Nature of Religion.* Boston, MA: Houghton Mifflin Harcourt, 1959.

Ellyatt, Holly. "'Tiger' of Eastern Europe Is Waking up: Romanian PM." *CNBC.Com.* November 2013.

Emporis. "Euro Tower." *Emporis.com.* 2020. https://www.emporis.com/buildings/309966/euro -tower-bucharest-romania.

Enache, Maria. "Metrorex vrea să rezilieze contractul cu Sindomet şi să scoată din staţii magazinele de textile şi covrigăriile." *MondoNews.Ro.* 2014. http://www.mondonews.ro /metrorex-vrea-sa-rezilieze-contractul-cu-sindomet-si-sa-scoata-din-statii-magazinele -de-textile-si-covrigariile/.

Endicott, John, Pamela Johnston, and Nancy F. Lin. *Underground Cities: New Frontiers in Urban Living.* London: Lund Humphries, 2020.

Engelke, Matthew. "Text and Performance in an African Church: The Book, 'Live and Direct.'" *American Ethnologist* 31, no. 1 (2004): 76–91.

Engels, Frederick. *The Housing Question.* Scotts Valley, CA: CreateSpace Independent Publishing Platform, 2016.

Epure, Ioana. "Am umblat prin Bucureşti să văd cât de penibilă e scumpirea tarifelor de parcare." *Vice.Com.* 2018.

Estrin, S. and M. Uvalic. "FDI into Transition Economies: Are the Balkans Different?" *Economics of Transition* 22, no. 2 (2014): 281–312.

"Eternitate dată în rate." *Evz.Ro.* 2006.

Etves, Ştefan. "Locurile de veci—Tot mai greu de găsit în capitală." *Gândul.Ro.* 2012. https:// www.gandul.ro/stiri/locurile-de-veci-tot-mai-greu-de-gasit-in-capitala-263353.

European Commission. "2021 Retrospective on Migration in Romania." https://ec.europa.eu /migrant-integration/news/2021-retrospective-migration-romania_en, April 3, 2022.

Eurostat. "Earnings Statistics." *ec.europa.eu.* 2015.

———. "Minimum Wage Statistics." *ec.europa.eu.* 2015. http://ec.europa.eu/eurostat/statistics -explained/index.php/Minimum_wage_statistics#Monthly_national_minimum_wages.

Farland, Maria. "Decomposing City: Walt Whitman's New York and the Science of Life and Death." *ELH* 74, no. 4 (2017): 799–827.

Fattal, Alexander L. *Guerrilla Marketing: Counterinsurgency and Capitalism in Colombia.* Chicago Studies in Practices of Meaning. Chicago: University of Chicago Press, 2018.

Featherstone, Mike. *Undoing Culture: Globalization, Postmodernism and Identity. British Journal of Sociology.* Published in Association with Theory, Culture & Society. New York: SAGE Publications, 1995.

Fehérváry, Krisztina. "American Kitchens, Luxury Bathrooms, and the Search for a 'Normal' Life in Postsocialist Hungary." *Ethnos: Journal of Anthropology* 67, no. 3 (November 2002): 369–400.

———. *Politics in Color and Concrete: Socialist Materialities and the Middle Class in Hungary.* Bloomington, IN: Indiana University Press, 2013.

Ferguson, James. *Expectations of Modernity: Myths and Meanings of Urban Life on the Zambian Copperbelt.* Berkeley: University of California Press, 1999.

———. *Global Shadows: Africa in the Neoliberal World Order.* Durham, NC: Duke University Press, 2006.

———. *Presence and Social Obligation: An Essay on the Share.* Chicago, IL: Prickly Paradigm Press, 2021.

Ferré-Sadurní, Luis et al. "How the Storm Turned Basement Apartments into Death Traps." *New York Times.* September 2, 2021. https://www.nytimes.com/2021/09/02/nyregion/basement-apartment-floods-deaths.html.

Filip, Paul. *Bellu: Panteon național.* 2nd-A ed. Bucharest: Editura Tradiție, 2001.

Fisch, Michael. *An Anthropology of the Machine: Tokyo's Commuter Train Network.* Chicago: University of Chicago Press, 2018.

Forbes.ro. "Biblioteca digitală reloaded: Peste 500 de titluri noi de cărți și audiobook-uri." *Forbes.Ro.* Bucharest, 2013. https://www.forbes.ro/biblioteca-digitala-reloaded-peste-500-de-titluri-noi-de-carti-si-audiobook-uri_0_6780-12051.

Foucault, Michel. *Abnormal: Lectures at the Collège de France, 1974–1975.* London: Picador, 2007.

———. "Of Other Spaces." *Diacritics* 16, no. 1 (1986): 22–27.

———. *Society Must Be Defended: Lectures at the College de France.* New York: Picador, 2003.

———. *The Birth of Biopolitics: Lectures at the Collège de France, 1978–79.* New York: Palgrave Macmillan, 2008.

———. "The Birth of Social Medicine." In *Power: The Essential Works of Foucault, 1954–1984, Vol. 3,* edited by James D. Faubion, 134–56. New York: New Press, 2000.

Frampton, Adam, Clara Wong, and Jonathan Solomon. *Cities Without Ground: A Hong Kong Guidebook.* Novato, CA: Oro Editions, 2012.

Frisby, David. *Cityscapes of Modernity: Critical Explorations.* Boston, MA: Polity Press, 2001.

———. "Analysing Modernity: Some Issues." In *Tracing Modernity: Manifestations of the Modern in Architecture and the City,* edited by Mari Hvattum and Christian Hermansen, 3–22. London: Taylor & Francis, 2004.

Gabor, Adrian. "Vodafone—Supernet City 4G, Internet Gratuit Si Garantarea Experientei Retelei." *Idevice.Ro.* 2016. https://www.idevice.ro/2016/02/08/vodafone-noutati-super-citi-4g/.

Gandy, Matthew. "The Paris Sewers and the Rationalization of Urban Space." *Transactions of the Institute of British Geographers* 24, no. 1 (1999): 23–44.

Garrett, Bradley L. *Bunker: What It Takes to Survive the Apocalypse.* New York: Scribner, 2021.

———. *Bunker: Building for the End Times.* New York: Penguin Books Limited, 2020.

Gavrilov, A. K. "Techniques of Reading in Classical Antiquity." *Classical Quarterly* 47, no. 1 (1997): 56–73.

Geană, Gheorghiță. "Remembering Ancestors: Commemorative Rituals and the Foundation of Historicity." *History and Anthropology* 16, no. 3 (September 1, 2005): 349–61.

Gheorghe, Ilinca. "Parcarea subterană de la Universitate a fost inaugurată | Nașul.TV." *Nașul. TV.* Accessed April 6, 2017. http://www.nasul.tv/parcarea-subterana-universitate-fost -inaugurata/.

Gheorghiu, Teodor Octavian. "Arhitectura subterană: Despre structurile urbane subterane medievale românești." *Arhitectura* 4 (1988): 43–45.

Ghica, Sorin. "Mafia de la metrou: Sindicalistul Ion Rădoi, ajutat de Mocanu și de Orban, face un munte de bani din chiriile luate pe spațiile comerciale." *Adevărul.Ro.* 2010. https:// adevarul.ro/news/eveniment/mafia-metrou-sindicalistul-ion-radoi-ajutat-mocanu -orban-munte-bani-chiriile-luate-spatiile-comerciale-1_50ae5e817c42d5a6639c1fa8 /index.html.

Ghodsee, Kristen and Mitchell Orenstein. *Taking Stock of Shock: Social Consequences of the 1989 Revolutions*. Oxford: Oxford University Press, 2021.

Gillet, Kit. "In Bucharest, the Old Town Sees New Life." *New York Times*. November 11, 2015. http://www.nytimes.com/2015/11/15/travel/bucharest-nightlife.html?_r=0.

Ginsburg, Faye. "Rethinking the Digital Age." In *Global Indigenous Media: Cultures, Poetics, and Politics*, edited by Pamela Wilson and Michelle Stewart, 287–305. Durham, NC: Duke University Press, 2008.

Giurescu, Dinu C. *Razing of Romania's Past*. New York: World Monuments Fund, 1989.

Gloviczki, Peter J. "Ceaușescu's Children: The Process of Democratization and the Plight of Romania's Orphans." *Critique* 3 (2004): 116–25.

Glynn, Tom. *Reading Publics: New York City's Public Libraries, 1754–1911*. New York: Fordham University Press, 2015.

Goffman, Erving. *The Presentation of Self in Everyday Life*. London: Overlook Press, 1973.

Gogu, Constantin Radu et al. "Urban Hydrogeology Studies in Bucharest City, Romania." *Procedia Engineering* 209 (2017): 135–42.

Gogu, Constantin Radu et al. "Urban Groundwater Modeling Scenarios to Simulate Bucharest City Lake Disturbance." *3rd Urban Hydrogeology Workshop: Current Trends and Approaches in Urban Hydrogeology* 53 (2015): 1–7.

Goldschmidt, Leopold A. "Air Rights." Chicago: American Society of Planning Officials. Information Report No. 186 1964. https://www.planning.org/pas/reports/report186.htm.

Gomez, James. "Europe's Hottest Building Market Needs Workers to Come Home." *Bloomberg.Com*. Prague, November 2018.

Graeber, David. *Direct Action: An Ethnography*. Oakland, CA: AK Press, 2009.

Graham, Stephen. "Luxified Skies: How Vertical Urban Housing Became an Elite Preserve." *City* 19, no. 5 (2015): 618–45.

———. *Vertical: The City from Satellites to Bunkers*. London: Verso Books, 2016.

Grama, Emanuela. *Socialist Heritage: The Politics of Past and Place in Romania*. Bloomington, IN: Indiana University Press, 2019.

Grant, Bruce. "The Edifice Complex: Architecture and the Political Life of Surplus in the New Baku." *Public Culture* 26, no. 3 (2014): 501–28.

Green, William and John F Doershuk. "Cultural Resource Management and American Archaeology." *Journal of Archaeological Research* 6, no. 2 (1998): 121–67.

Greenberg, Jessica. "On the Road to Normal: Negotiating Agency and State Sovereignty in Post-socialist Serbia." *American Anthropologist* 113, no. 1 (2011): 88–100.

Groot, Margo and Teresa Hackett. "Through the PULMAN Glass: Looking at the Future of Libraries in Europe." *New Library World* 104, no. 3 (2003): 103–9.

Gupta, Akhil and James Ferguson. "Discipline and Practice: 'The Field' as Site, Method, and Location in Anthropology." In *Anthropological Locations: Boundaries and Grounds of a Field Science*, edited by Akhil Gupta and James Ferguson, 1–46. Berkeley: University of California Press, 1997.

Habermas, Jürgen. *The Structural Transformation of the Public Sphere: An Inquiry into a Category of Bourgeois Society.* London: Wiley, 2015.

Harris, Andrew. "Vertical Urbanisms: Opening Up Geographies of the Three-Dimensional City." *Progress in Human Geography* 39, no. 5 (2015): 601–20.

Harris, Nigel. "Structural Adjustment and Romania." *Economic and Political Weekly* 29, no. 44 (October 1994): 2861–64.

Harvey, David. "Flexible Accumulation through Urbanization: Reflections on 'Post-Modernism' in the American City." *Antipode* 19, no. 3 (1987): 260–86.

Haslam, Dave. *Life After Dark: A History of British Nightclubs & Music Venues.* London: Simon & Schuster U.K., 2015.

Hatherley, Owen. *Landscapes of Communism: A History Through Buildings.* New York: New Press, 2016.

Hayek, F A. *The Road to Serfdom.* Routledge Classics. London: Taylor & Francis, 1976.

Heap, Chad. *Slumming: Sexual and Racial Encounters in American Nightlife, 1885–1940.* Chicago: University of Chicago Press, 2008.

Heidegger, Martin. *The Fundamental Concepts of Metaphysics: World, Finitude, Solitude.* Bloomington, IN: Indiana University Press, 2001.

Hénard, Eugène. "The Cities of the Future." *Transactions: The Royal Institute of British Architects* (1911): 345–67. http://urbanplanning.library.cornell.edu/DOCS/henard.htm.

Henderson, Jason. "The Spaces of Parking: Mapping the Politics of Mobility in San Francisco." *Antipode* 41, no. 1 (January 2009): 70–91.

Hewitt, Lucy and Stephen Graham. "Vertical Cities: Representations of Urban Verticality in 20th-Century Science Fiction Literature." *Urban Studies* 52, no. 5 (2015): 923–37.

Ho, Karen. *Liquidated: An Ethnography of Wall Street.* Durham, NC: Duke University Press Books, 2009.

Hodder, Ian. *Towards Reflexive Method in Archaeology: The Example at Çatalhöyük.* Cambridge: British Institute at Ankara, 2000.

Hohne, Stefan. *Riding the New York Subway: The Invention of the Modern Passenger.* Cambridge, MA: MIT Press, 2021.

Hood, Clifton. *722 Miles: The Building of the Subways and How They Transformed New York.* Baltimore, MD: Johns Hopkins University Press, 2004.

Horst, Heather and Daniel Miller. *The Cell Phone: An Anthropology of Communication.* Oxford: Berg Publishers, 2006.

Iancu, Liviu. "Peste 200.000 de mașini vechi sunt abandonate în parcările din București." *Ziarul Financiar.* 2014. http://www.zf.ro/zf-24/peste-200-000-de-masini-vechi-sunt-aband onate-in-parcarile-din-bucuresti-12517902.

Iancu, Marcel. "Utopia Bucureștilor." In *Către o arhitectură a Bucureștilor*, edited by Marcel Iancu, Horia Creangă, and Octav Doicescu, 7–20. Bucharest: Tribuna Edilitară, 1935.

Ignat, Vlad. "Păianjenul Metrorex schimbă harta capitalei. Bucureştenii se mută în subteran, din 2030." *Adevărul.Ro.* August 28, 2012. http://adevarul.ro/news/bucuresti/paianjenul -metrorex-schimba-harta-capitalei-bucurestenii-muta-subteran-2030-1_50bdec7b7c 42d5a663d01c96/index.html.

———. "Zeci de şoferi ignoră parcarea subterană de la Universitate şi îşi lasă maşinile în zona pietonală." *Adevărul.Ro.* 2013. http://adevarul.ro/news/bucuresti/zeci-soferi-ignora -parcarea-subterana-universitate-isi-lasa-masinile-zona-pietonala-1_510ac8f34b 62ed5875c1335a/index.html.

IGSU. "Situaţia cu coordonatele GPS ale adăpostruilor publice de protecţie civilă din munici- piul Bucureşti." Bucharest, 2020. https://www.primariasector1.ro/download/situatii-de -urgenta/IGSU—ADĂPOSTURI PUBLICE- 2020.pdf.

Ilie, Cornel-Constantin. *Între Bellu şi Montparnasse. Români celebri în faţa morţii.* Bucharest: Editura Historia, 2008.

"Incendiu în clubul Colectiv. Autorităţile cer cetăţenilor să doneze sânge etapizat, iar de donare de piele nu e nevoie deocamdată." *Adevărul.Ro.* 2015. http://adevarul.ro/news/eveniment /autoritatile-cer-cetatenilor-doneze-sange-etapizat-donare-piele-nu-e-nevoie-deocam data-1_56363f1ff5eaafab2c37a33e/index.html.

Interparking Group. "Interparking—Activity Report 2017." Brussels, 2018.

———. "Interparking Activity Report 2012." Brussels, 2013.

———. "Piaţa Universităţii (Bucharest)." https://www.interparking-romania.ro, 2019.

Iolu, Liviu. "O traumă naţională." *Adevărul.Ro.* 2015. http://adevarul.ro/news/eveniment/o -trauma-nationala-nu-mai-stinsa-1_56362ad3f5eaafab2c3737c9/index.html.

Ion, Elena. "Public Funding and Urban Governance in Contemporary Romania: The Resur- gence of State-Led Urban Development in an Era of Crisis." *Cambridge Journal of Regions, Economy and Society* 7, no 1 (2014): 171–187.

Ionescu, Doru. *CLUB A: 42 de ani.* Bucharest: Casa de Pariuri Literare. 2011.

ITA-AITES. "Urban Underground Space in a Changing World: Delivering; Planning; Decid- ing." ITA Global Perspective Programme (#01). Lausanne, (2012): 1–11.

Ivan, Daniela. "Interviu Noica—Blocurile turn, un pericol seismic." *Jurnalul.Ro.* 2006. https:// jurnalul.antena3.ro/special-jurnalul/interviuri/interviu-noica-blocurile-turn-un -pericol-seismic-18138.html.

Ivanov, Catiuşa. "Metrorex a rupt contractul pentru spaţiile publicitare de la metrou cu firma sindicatelor. Redevenţa pentru spaţiile comerciale va fi renegociată." *HotNews.Ro.* 2014. https://www.hotnews.ro/stiri-administratie_locala-16857656-metrorex-rupt-contractul -pentru-spatiile-publicitare-metrou-firma-sindicatelor-redeventa-pentru-spatiile -comerciale-renegociata.htm.

———. "Piaţa locurilor de veci prosperă şi pe vreme de criză." *HotNews.Ro.* https://www .hotnews.ro/stiri-esential-6037641-piata-locurilor-veci-prospera-vreme-criza.htm.

———. "Preţul plătit pentru dezvoltarea haotică a Bucureştiului: Lacul din Parcul Circului aproape a secat, habitatul superbilor lotuşi egipteni fiind în pericol.." *HotNews.Ro.* 2019. https://www.hotnews.ro/stiri-administratie_locala-23135354-foto-pretul-platit-pentru -dezvoltarea-haotica-bucurestiului-lacul-din-parcul-circului-aproape-secat-habitatul -superbilor-lotusi-egipteni-fiind-pericol.htm.

"În ce locuri ne putem adăposti în situaţii de urgenţă?—Infografic adăposturi civile din Bucureşti." *Metropotam.* 2015. http://metropotam.ro/La-zi/unde-ne-putem-adaposti-in -situatii-de-urgenta-infografic-adaposturi-civile-art2076720958/.

Jackson, Kenneth T. *Crabgrass Frontier: The Suburbanization of the United States.* Oxford: Oxford University Press, 1985.

Jacobs, Jane. *The Death and Life of Great American Cities.* New York: Vintage Books, 1961.

Jameson, Fredric. *Postmodernism, Or, The Cultural Logic of Late Capitalism.* Durham, NC: Duke University Press, 1991.

Jastrzab, Mariusz. "Cars as Favors in People's Poland." In *The Socialist Car: Automobility in the Eastern Bloc,* edited by Lewis H. Siegelbaum. Ithaca, NY: Cornell University Press, 2011.

Jelavich, Peter. *Berlin Cabaret.* Cambridge, MA: Harvard University Press, 1996.

Johnson, Peter. "The Modern Cemetery: A Design for Life." *Social & Cultural Geography* 9, no. 7 (2008): 777–90.

———. "Unravelling Foucault's 'Different Spaces.'" *History of the Human Sciences* 19, no. 4 (2006): 75–90.

Johnson, W. *The Broken Heart of America: St. Louis and the Violent History of the United States.* New York: Basic Books, 2020.

Jose, George. "Hawa Khaana in Vasai Virar: Urban Housing and the Commodification of Airspace in Mumbai's Periphery." *City* 21, no. 5 (2017): 632–40.

Joyce, Patrick. *The Rule of Freedom: Liberalism and the Modern City.* London: Verso, 2003.

Lin, G., S. P. Luo, and J. Ni. "Damages of Metro Structures Due to Earthquake and Corresponding Treatment Measures." *Modern Tunnelling Technology* 46, no. 4 (2009): 3–6.

Juris, Jeffrey S. *Networking Futures: The Movements against Corporate Globalization.* Durham, NC: Duke University Press, 2008.

Kaparosa, George and Pantoleon Skayannisb. "Dealing with Context and Uncertainty in the Development of the Athens Metro Base Project." *Planning Theory and Practice* 15, no. 3 (2014): 403–9.

Karasz, Palko. "At Ceaușescu's Villa, Focus is on Décor, Not Dictatorship." *New York Times.* June 6, 2016. https://www.nytimes.com/2016/06/07/world/europe/romania-bucharet-ceausescu-villa.html.

Kern, Rebecca. "It's a Cellars Market." *Washington Post.* January 18, 2013. https://www.washingtonpost.com/express/wp/2013/01/18/its-a-cellars-market/.

Killada, Mohana R. and G. V. R. Raju. "World's Top Economies and Their Metro Systems' Ridership and Financial Performance." *International Journal of Traffic and Transportation Engineering* 7, no. 4 (2018): 91–97.

Kim, Annette M. "The Extreme Primacy of Location: Beijing's Underground Rental Housing Market." *Cities* 52 (2016): 148–58

Kligman, Gail. *The Politics of Duplicity: Controlling Reproduction in Ceaușescu's Romania.* Contraversions, Critical Studies in Jewish Literature Culture and Society, Vol. 11. Berkeley: University of California Press, 1998.

———. *The Wedding of the Dead: Ritual, Poetics, and Popular Culture in Transylvania.* Berkeley: University of California Press, 1988.

Klinenberg, Eric. *Palaces for the People: How Social Infrastructure Can Help Fight Inequality, Polarization, and the Decline of Civic Life.* New York: Crown, 2018.

Koblenko, Mihai. "Cât te costă un buncăr antiatomic și ce poți face în acest refugiu subteran ca să nu te plictisești." *Național.* 2015. https://www.national.ro/old/cat-te-costa-un-buncar-antiatomic-si-ce-poti-face-in-acest-refugiu-subteran-ca-sa-nu-te-plictisesti-463463.html.

Kociatkiewicz, Jerzy and Monika Kostera. "The Anthropology of Empty Spaces." *Qualitative Sociology* 22, no. 1 (1999): 37–50.

Kuznetsov, Sergey et al. *Hidden Urbanism: Architecture and Design of the Moscow Metro, 1935–2015*. Berlin: DOM Publishers, 2016.

"Lacul Circului din București seacă văzând cu ochii din cauza construcției unui bloc." *Antena3.Ro*. 2009. https://www.antena3.ro/actualitate/lacul-circului-din-bucuresti-seaca -vazand-cu-ochii-din-cauza-constructiei-unui-bloc-68623.html.

Lancione, Michele. "The Politics of Embodied Urban Precarity: Roma People and the Fight for Housing in Bucharest, Romania." *Geoforum* 101. (2019): 182–91.

——. "Underground Inscriptions." *Cultural Anthropology* 35, no. 1 (2020): 31–39.

——. "Weird Exoskeletons: Propositional Politics and the Making of Home in Underground Bucharest." *International Journal of Urban and Regional Research* 43, no. 3 (May 4, 2019): 535–50.

Lanzano, G., E. Bilotta, and G. Russo. "Tunnels under Seismic Loading: A Review of Damage Case Histories and Protection Methods." In *Strategies for Reduction of the Seismic Risk*, edited by Giovanni Fabbrocino and Filippo Santucci de Magistris, 65–74. Ripalimosani: Arti Grafiche La Regione srl, 2008.

Lavinia, Stan. "Romanian Privatization: Assessment of the First Five Years." *Communist and Post-Communist Studies* 28, no. 4 (December 1995): 427–35.

Lefebvre, Henri. *Rhythmanalysis: Space, Time and Everyday Life*. New York: Bloomsbury Publishing, 2008.

——. *The Production of Space*. London: Wiley-Blackwell, 1992.

——. *Writings on Cities*. London: Wiley-Blackwell, 1996.

Leitner, Helga et al. "Space Grabs: Colonizing the Vertical City." *International Journal of Urban and Regional Research* (2020): 1–11.

Lemon, Alaina. "MetroDogs: The Heart in the Machine." *Journal of the Royal Anthropological Institute* 21, no. 3 (2015): 660–79.

——. "Talking Transit and Spectating Transition: The Moscow Metro." In *Altering States: Ethnographies of Transition in Eastern Europe and the Former Soviet Union*, edited by Daphne Berdahl, Matti Bunzl, and Martha Lampland, 14–39. Ann Arbor: University of Michigan Press, 2000.

Leshem, Noam. "'Over Our Dead Bodies': Placing Necropolitical Activism." *Political Geography* 45 (2015): 34–44.

Liao, Ran. "Overview of Anti-Seismic Researches of Underground Structures." *E3S Web of Conferences* 38 (2018): 3–6.

Lică, Cristina. "www.Cimitir.Ro: Metrul pătrat de țărână, cât cel de apartament. Iar locul de veci nu e pe veci." *Evz.Ro*. 2012. https://evz.ro/wwwcimitirro-metrul-patrat-de-tarana-cat -cel-de-apartament-973663.html.

Low, Setha M. *Behind the Gates: Life, Security, and the Pursuit of Happiness in Fortress America*. New York: Routledge, 2004.

Lowe, David and Tony Joel. *Remembering the Cold War: Global Contest and National Stories*. London: Taylor & Francis, 2014.

MacGaffey, Janet and Rémy Bazenguissa-Ganga. *Congo-Paris: Transnational Traders on the Margins of the Law*. London: International African Institute, 2000..

Machedon, Luminița and Ernie Scoffham. *Romanian Modernism: The Architecture of Bucharest 1920–1940*. Cambridge, MA: MIT Press, 1999.

"Mafia mormintelor de la Bellu: Samsarii promit că scot morții din groapă și cer până la 8.000 de euro pentru un loc de veci!" *Adevărul.Ro*. 2011. https://adevarul.ro/news/bucuresti

/exclusiv-mafia-mormintelor-bellu-samsarii-promit-scot-mortii-groapa-cer-8000-euro
-loc-veci-1_50bde9767c42d5a663cfd1f1/index.html.

Makovicky, Nicolette. *Neoliberalism, Personhood, and Postsocialism: Enterprising Selves in
Changing Economies.* London: Taylor & Francis, 2016.

Malinowski, Bronisław. *Argonauts of the Western Pacific: An Account of Native Enterprise and
Adventure in the Archipelagoes of Melanesian New Guinea.* New York: Taylor and Francis, 2003.

Malkki, Liisa H. "News and Culture: Transitory Phenomena and the Fieldwork Tradition." In
Anthropological Locations: Boundaries and Grounds of a Field Science, edited by Akhil
Gupta and James Ferguson, 86–101. Berkeley: University of California Press, 1997.

Mao, Ruichang et al. "Global Urban Subway Development, Construction Material Stocks, and
Embodied Carbon Emissions." *Humanities and Social Sciences Communications* 8, no. 1
(2021): 83.

Marcuse, Peter. "Gentrification, Abandonment, and Displacement: Connections, Causes, and
Policy Responses in New York City." *Washington University Journal of Urban and Contemporary Law* 28 (January 1985): 195–240.

Marea Adunare Națională. Legea nr. 2/1978 privind apărarea civilă în Republica Socialistă
Română. Bucharest, 1978.

Marin, Iulia. "Sorin Oprescu inaugurează joi parcarea subterană de la Universitate." *Adevărul.
Ro.* 2012. http://adevarul.ro/news/eveniment/sorin-oprescu-inaugureaza-joi-parcarea—
subterana-universitate-1_50a46e8a7c42d5a66368c8be/index.html.

Martin, Emily. "Flexible Survivors." *Cultural Values* 4, no. 4 (2000): 512–17..

Masco, Joseph. "Life Underground: Building a Bunker Society." *Anthropology Now* 1, no. 2
(2009): 13–29.

———. *The Future of Fallout, and Other Episodes in Radioactive World-Making.* Durham, NC:
Duke University Press, 2020.

Massumi, Brian. *A User's Guide to Capitalism and Schizophrenia: Deviations from Deleuze and
Guattari.* Cambridge, MA: MIT Press, 1992.

Măgureanu, Andrei and Despina Măgureanu. "Preventive Archaeology in Romania Between
Negotiation and Myth: Some Thoughts." In *Recent Developments in Preventive Archaeol-
ogy in Europe,* edited by Predrag Novaković et al., 257–70. Ljubljana: Ljubljana University
Press, 2016.

"Măsuri în contra atacurilor aeriene." *Adevărul.Ro.* August 15, 1916.

Mbembe, Achille. *Necropolitics.* Durham, NC: Duke University Press, 2019.

MCCANN Worldgroup. "Case Study: Vodafone the Digital Library by McCann Bucharest."
youtube.com. 2013. https://www.youtube.com/watch?v=u1UMKGi9nuw.

———. "Vodafone: Digital Library Wallpaper." *youtube.com.* 2014. https://www.youtube.com
/watch?v=E01XZyVsWJY.

"McDonald's a închis restaurantul din pasajul de la Unirii, după 20 de ani: Decizia nu ne
aparține." *Adevărul.Ro.* 2019. https://adevarul.ro/stiri-locale/bucuresti/mcdonalds-a
-inchis-restaurantul-din-pasajul-de-la-1944949.html.

McIntosh, Janet. "Mobile Phones and Mipoho's Prophecy: The Powers and Dangers of Flying
Language." *American Ethnologist* 37, no. 2 (2010): 337–53.

McKinlay, Doug. "Switzerland's Null Stern Hotel: The Nuclear Option." *Guardian.* September 12, 2009. https://www.theguardian.com/travel/2009/sep/13/nuclear-bunker-hotel-null
-stern-switzerland.

McManners, John. *Death and the Enlightenment: Changing Attitudes to Death Among Christians and Unbelievers in Eighteenth-Century France*. Oxford: Oxford University Press, 1985.

McNeil, Donald G. "Bucharest Journal; In a City Gone to the Dogs, Bardot's on the Case." *New York Times*. March 2, 2001. https://www.nytimes.com/2001/03/02/world/bucharest-journal-in-a-city-gone-to-the-dogs-bardot-s-on-the-case.html.

McNeill, Donald. "Volumetric Urbanism: The Production and Extraction of Singaporean Territory." *Environment and Planning A* 51, no. 4 (2019): 849–68.

MDRAP. "Norme tehnice privind proiectarea şi executarea adăposturilor şi a punctelor de comandă de protecţie civilă." Bucharest, 2016. https://www.oar-bucuresti.ro/anunturi/2016/02/26/c/16-02-25 Proiect Normativ adaposturi.pdf.

Melly, Caroline. *Bottleneck: Moving, Building, and Belonging in an African City*. Chicago: University of Chicago Press, 2017.

Mendelski, Martin. "15 Years of Anti-Corruption in Romania: Augmentation, Aberration and Acceleration." *European Politics and Society* 22, no. 2 (2021): 237–58.

Mendez, Pablo and Noah Quastel. "Subterranean Commodification: Informal Housing and the Legalization of Basement Suites in Vancouver from 1928 to 2009." *International Journal of Urban and Regional Research* 39, no. 6 (2016): 1155–71.

Meskell, Lynn. *The Nature of Heritage: The New South Africa*. London: Wiley, 2011.

"Metrorex reduce la jumătate tariful în parcarea de la Străuleşti." *Digi24.Ro*. 2019. https://www.digi24.ro/stiri/economie/transporturi/metrorex-reduce-la-jumatate-tariful-in-parcarea-de-la-straulesti-1073545.

Meyer, Hanns. *In the Good Ratskeller of Bremen: A Contribution to the History of German Innkeeping*. H. M. Hauschild, Bremen: 1959.

Meyer, Michael. "Moment of Truth." *Newsweek*. March 2005. https://www.newsweek.com/moment-truth-114327.

Miao, Yu et al. "Seismic Response of a Subway Station in Soft Soil Considering the Structure-Soil-Structure Interaction." *Tunnelling and Underground Space Technology* 106, no. October (2020): 103629.

"Mii de maşini vechi abandonate în Bucureşti, în ciuda crizei de locuri de parcare. Ce riscă proprietarii lor." *Stirileprotv.Ro*. 2016.

Miller, Claire Cain. "Where Young College Graduates Are Choosing to Live." *New York Times*. October 20, 2014. http://www.nytimes.com/2014/10/20/upshot/where-young-college-graduates-are-choosing-to-live.html?_r=0.

Miller, David S. *101 Places to Get F*cked Up Before You Die: The Ultimate Travel Guide to Partying Around the World*. New York: St. Martin's Press, 2014.

Miller, Peter and Nikolas Rose. "Mobilizing the Consumer: Assembling the Subject of Consumption." *Theory, Culture & Society* 14, no. 1 (1997): 1–36.

Mironescu, Vlad. "40 de schelete umane au fost descoperite în piaţa Universităţii." *Gândul. Ro*. 2012. https://www.gandul.ro/stiri/40-de-schelete-umane-au-fost-descoperite-in-piata-universitatii-galerie-foto-7755188.

Moga, Cristi. "Mall-urile au creat 20.000 de locuri de parcare în Bucureşti. Primăria încă face parcări pe hârtie." *Ziarul Financiar*. 2009. http://www.zf.ro/eveniment/mallurile-au-creat-20-000-de-locuri-de-parcare-in-bucuresti-primaria-inca-face-parcari-pe-hartie-5048905/.

Morgan, Diane. "Bunker Conversion and the Overcoming of Siege Mentality." *Textual Practice* 31, no. 7 (November 10, 2017): 1333–60.

Morrison, Lynn. "Ceaușescu's Legacy: Family Struggles and Institutionalization of Children in Romania." *Journal of Family History* 29, no. 2 (April 1, 2004): 168–82.

Mount10 Company. "Mount10." *https://mount10.com/en/*. 2021. https://mount10.com/en/#video -6-sm.

Mumford, Lewis. *Technics and Civilization*. Chicago: University of Chicago Press, 2010.

———. *The Culture of Cities*. New York: Open Road Media, 1938.

Murray, Martin. *City of Extremes: The Spatial Politics of Johannesburg*. Durham, NC: Duke University Press, 2011.

Musteață, Sergiu. "Preserving Archaeological Remains in Situ: From the Legal to the Practical Issues. The Romanian Case." In *Current Trends in Archaeological Heritage Preservation: National and International Perspectives*, edited by Sergiu Musteață and Ștefan Caliniuc, 15–20. Iași: BAR International, 2015.

Myers, Steven Lee. "'Please Save Us!' Grim Scenes in China as Flood Inundates a Subway." *New York Times*. July 20, 2021. https://www.nytimes.com/2021/07/20/world/asia/china-flooding -zhengzhou-subway.html?referringSource=articleShare.

Myradl, Gunnar. *Challenge to Affluence*. New York: Pantheon Books, 1965.

Nabas, Bushra. "The Role of Uncomfortable Heritage in Sustainable Development." *International Journal of Heritage and Museum Studies* 1, no. 1 (2019): 57–79.

Nadkarni, Maya. *Remains of Socialism: Memory and the Futures of the Past in Postsocialist Hungary*. Ithaca, NY: Cornell University Press, 2020.

Nalewicki, Jennifer. "Switzerland's Historic Bunkers Get a New Lease on Life." *Smithsonian Magazine*. Washington, D.C., March 2016. https://www.smithsonianmag.com/travel /switzerlands-bunkers-get-new-lease-life-180958233/.

Negru, Natalia. *De la stradă la ansambluri rezidențiale. Opt ipostaze ale locuirii în Bucureștiul contemporan*. Bucharest: Editura Pro Universitaria, 2016.

Nelson, Charles A. *Romania's Abandoned Children*. Cambridge, MA: Harvard University Press, 2014.

Nemeti, Sorin. "Manifesto for the Romanian Public Archaeology." *Journal of Ancient History and Archaeology* 4, no. 3 (2017): 5–7.

Novitchi, Ioan. "Metroul din București." *Arhitectura* 4 (1980): 11.

O'Neill, Bruce. "Corruption Kills." Edited by Dace Dzenovska and Larisa Kurtović. Hot Spots, Fieldsights. *CultAnth.org*. 2018. https://culanth.org/fieldsights/corruption-kills.

———. "The Ethnographic Negative: Capturing the Impress of Boredom and Inactivity." *Focaal: Journal of Global and Historical Anthropology* 78 (2017): 23–37.

———. "The Political Agency of Cityscapes: Spatializing Governance in Ceaușescu's Bucharest." *Journal of Social Archaeology* 9, no. 1 (February 2009): 92–109.

———. *The Space of Boredom: Homelessness in the Slowing Global Order*. Durham, NC: Duke University Press, 2017.

———. "Up, Down, and Away: Placing Privilege in Bucharest, Romania." *Journal of the Royal Anthropological Institute* 28, no. 1 (2022): 130–51.

O'Neill, Kevin Lewis, and Benjamin Fogarty-Valenzuela. "Verticality." *Journal of the Royal Anthropological Institute* 19, no. 2 (June 1, 2013): 378–89.

Oberlander-Tarnoveanu, Irina. "Preventive Archaeological Research in Romania—Legal Aspects and Results Dissemination." In *European Preventive Archaeology Papers of the EPAC Meeting, Vilnius 2004*, edited by Katalin Bozóki-Ernyey, 167–79. Budapest: National Office of Cultural Heritage, Hungary–Council of Europe, 2007.

Ochiană, Mihaela. "O altă utilitate pentru beciuri și subsoluri: Pot fi transformate în buncăre antiatomice, ieftin și ușor." *Imopedia.Ro.* 2015. https://media.imopedia.ro/stiri-imobiliare /buncare-antiatomice-beci-subsol-23393.html.

Ofițeru, Andreea. "Parcarea subterană de la Universitate." *Gândul.Ro.* 2012. http://www.gandul .info/stiri/parcarea-subterana-de-la-universitate-cum-arata-si . . . -de-parcare-vezi -harta-cu-strazile-unde-nu-mai-poti-parca-galerie-foto-10324330.

Ong, Aihwa. "Graduated Sovereignty in South-East Asia." *Theory, Culture & Society* 17, no. 4 (August 2000): 55–75. http://tcs.sagepub.com/content/17/4/55.abstract.

———. "Hyperbuilding: Spectacle, Speculation, and the Hyperspace of Sovereignty." In *Worlding Cities: Asian Experiments and the Art of Being Global*, edited by Ananya Roy and Aihwa Ong, 205–26. Chichester: Blackwell, 2011.

Onwuemezi, Natasha. "Shakespeare 'Digital Library Wallpaper' from British Library and Vodafone." *Bookseller.* April 26, 2016. https://www.thebookseller.com/news/digital-library -wallpaper-327630.

Oprescu, Sorin, ed. *Bellu: The Garden of Souls.* Bucharest: Primăria Municipiului, 2013.

Osborne, Thomas and Nikolas Rose. "Governing Cities: Notes on the Spatialisation of Virtue." *Environment and Planning D: Society and Space* 17, no. 6 (1999): 737–60.

Ost, David. "Stuck in the Past and the Future: Class Analysis in Postcommunist Poland." *East European Politics and Societies* 29, no. 3 (August 1, 2015): 610–24.

"Parcare sau sit arheologic?" *Rfi.Ro.* 2010. https://www.rfi.ro/articol/stiri/economie/parcare-sit -arheologic.

Park, Robert E. and Ernest W. Burgess. *The City.* Chicago: University of Chicago Press, 2012.

Parlamentul României. Hotărârea nr. 560/2005 pentru aprobarea categoriilor de construcții la care este obligatorie realizarea adăposturilor de protecție civilă, precum și a celor la care se amenajează puncte de comandă (2005).

———. Lege Nr. 481/2004 (2004).

———. Legea protecției civile nr. 106/1996 (1996).

Pârvulescu, Diana. "Noaptea Muzeelor: 25.000 de vizitatori la cimitirul Bellu din capitală." *MediaFax.Ro.* 2015. https://www.mediafax.ro/cultura-media/noaptea-muzeelor-25-000-de -vizitatori-la-cimitirul-bellu-din-capitala-foto-14280827.

Passaro, Joanne. "'You Can't Take the Subway to the Field!': 'Village' Epistemologies in the Global Village." In *Anthropological Locations: Boundaries and Grounds of a Field Science*, edited by Akhil Gupta and James Ferguson, 147–62. Berkeley: University of California Press, 1997.

Patico, Jennifer and Melissa L Caldwell. "Consumers Exiting Socialism: Ethnographic Perspectives on Daily Life in Post-Communist Europe." *Ethnos: Journal of Anthropology* 67, no. 3 (January 1, 2002): 285–94.

Patron, Luminița et al. "Conceptul strategic București 2035." Bucharest, 2010.

Pen, Cees-Jan and Marloes Hoogerbrugge. "Economic Vitality of Bucharest." The Hague, 2012.

Perec, Georges. *An Attempt at Exhausting a Place in Paris.* Adelaide: Wakefield Press, 2010.

"Pericol de surpare în zona Eroilor. Traficul rutier rămâne închis 30 de zile." *Digi24.Ro.* 2015. https://www.digi24.ro/stiri/actualitate/evenimente/pericol-de-surpare-in-zona-eroilor -traficul-rutier-ramane-inchis-30-de-zile-466982.

Perrin, Richard. "Romania as the Destination for SSCs and BPO." Bucharest, 2012.

Peteri, Gyorgy. "Alternative Modernity? Everyday Practices of Elite Mobility in Communist Hungary, 1956–1980." In *The Socialist Car: Automobility in the Eastern Bloc*, edited by Lewis H Siegelbaum, 47–68. Ithaca, NY: Cornell University Press, 2011.

Petrescu, Dan. "Romania Country Brief: Europe and Central Asia Region." Bucharest: World Bank, 2002. http://lnweb90.worldbank.org/eca/eca.nsf/0/c4cfb7b8c4d1658185256c240050 a6a4/$FILE/Romania Country Brief.pdf.

Pike, David L. "Cold War Reduction: The Principle of the Swiss Bunker Fantasy." *Space and Culture* 20, no. 1 (2017): 94–106.

———. *Metropolis on the Styx: The Underworlds of Modern Urban Culture, 1800–2001*. Ithaca, NY: Cornell University Press, 2007.

———. *Subterranean Cities: The World Beneath Paris and London, 1800–1945*. Ithaca, NY: Cornell University Press, 2005.

Piketty, Thomas. *Capital in the Twenty-First Century*. Cambridge, MA: Harvard University Press, 2014.

Pine, Jason. "Economy of Speed: The New Narco-Capitalism." *Public Culture* 19, no. 2 (2007): 357–66.

PMB. "Strategia de parcare pe teritoriul municipiului Bucureşti." Bucharest, 2008.

Poggiali, Lisa. "Digital Futures and Analogue Pasts?: Citizenship and Ethnicity in Techno-Utopian Kenya." *Africa: The Journal of the International African Institute* 87, no. 2 (2017): 253–77.

Popescu, Adam. "Norii ruginii peste casele de lux din sud | VIDEO." *Evz.Ro*. 2008. https://evz .ro/nori-ruginii-peste-casele-de-lux-din-sud-video-809214.html.

Popescu, Mihai and Andrei Feraru. "Convorbire cu . . . Conf. Arh. Traian Stănescu." *Arhitectura* 4 (1980): 12–13.

"Povestea lui Bezviconi, istoricul nevoit să lucreze ca portar la cimitirul Bellu." *https:// bercenidepoveste.ro*. 2016. https://bercenidepoveste.ro/portar-la-cimitirul-bellu/.

Preda, Ionut et al. "Incendiu fără precedent în club Colectiv din Bucureşti: Cel puţin 27 de morţi şi 146 de persoane internate. A fost activat codul roşu de urgenţă." *Adevărul.Ro*. 2015. http://adevarul.ro/news/eveniment/explozie-clubul-colectiv-capitala-zeci-raniti-1 _5633e09ff5eaafab2c2b7df0/index.html.

Preda, Ionut and Anca Vancu. "Reacţii oficiale după incendiul de la club Colectiv. Klaus Iohannis: 'Nu mai putem continua pe principiul, «Lasă că merge şi aşa».'" *Adevărul.Ro*. 2015. http://adevarul.ro/news/eveniment/reactii-oficiale-incendiul-club-colectiv-dragnea -tineri-plecat-asculte-muzica-nu-mai-sunt-1_56344653f5eaafab2c2d7228/index.html.

Predescu, Alexandru. *Vremuri vechi bucureştene*. Bucharest: Editura Pentru Turism, 1990.

Quayson, Ato. *Oxford Street, Accra: City Life and the Itineraries of Transnationalism*. Durham, NC: Duke University Press, 2014.

"Probleme la metrou, pe magistrala Berceni—Pipera. Aglomeraţie la staţia Aviatorilor." *Adevărul.Ro*. 2017. https://adevarul.ro/news/bucuresti/probleme-metrou-magistrala -berceni-pipera-aglomeratie-statia-aviatorilor-1_58bd06ea5ab6550cb810d6fb/index .html

Racu, Radu. "Cât costă să faci din centrul Bucureştiului un al doilea Sibiu." *Ziarului Financiar*. 2011. http://www.zf.ro/business-construct/cat-costa-sa-faci-din-centrul-bucurestiului -un-al-doilea-sibiu-8281928.

Radu, Cristina. "Înfiinţarea autorităţii de transport metropolitan, benzi de circulaţie pentru transportul public, construcţia de parcări la intrările în oraş, piste de biciclete, prevăzute în planul de mobilitate urbană Bucureşti-Ilfov Aprobat de CGMB." *News.Ro.* 2017. https:// www.news.ro/social/infiintarea-autoritatii-transport-metropolitan-benzi-circulatie -transportul-public-constructia-parcari-intrarile-oras-piste-biciclete-prevazute-planul -mobilitate-urbana-bucuresti-ilfov-aprobat-cgmb-19224031290020170314 16875878.

Răduţă, Cristina. "A fost reparată avaria de la canalul colector de lângă podul Izvor." *Adevărul. Ro.* 2015. https://adevarul.ro/news/bucuresti/foto-fost-reparata-avaria-canalul-colector -podul-izvor-1_55603b90cfbe376e35804f52/index.html.

———. "Adevărul despre tunelurile subterane ale Bucureştiului: Care era rolul lor iniţial şi pe unde plănuia să scape Nicolae Ceauşescu la nevoie." *Adevărul.Ro.* August 30, 2015. https:// adevarul.ro/news/bucuresti/adevarul-despre-tunelurile-subterane-bucurestiului-era -rolul-initial-planuia-scape-nicolae-ceausescu-nevoie-1_55e2f3d1f5eaafab2c111c17/index .html.

———. "Cum a apărut o groapă mare de 8 metri şi adâncă de 4 metri lână podul Izvor. Un canal vechi de 104 ani, ignorat complet de toţi edilii capitalei, a surpat pământul." *Adevărul. Ro.* 2015. https://adevarul.ro/news/bucuresti/cum-aparut-groapa-mare-8-metri-adanca -4-metri-podul-izvor-canal-vechi-104-ani-ignorat-complet-edilii-capitalei-surpat -pamantul-1_555c6699cfbe376e35671d16/index.html.

———. "Metrorex a rupt contractul cu Sindomet pentru spaţiile publicitare, iar buticurile cu ojă şi lenjerie intimă de la metrou ar putea fi desfiinţate." *Adevărul.Ro.* 2014.

———. "Parcarea de la Intercontinental." *Adevărul.Ro.* 2016. http://adevarul.ro/news/bucu resti/parcarea-intercontinental-gata-octombrie-3-milioane-euro-pretul-partea-cons tructie-1_53b3e8990d133766a812e325/index.html..

———. "S-a rupt asfaltul în zona Tineretului. Apa Nova: „Este o avarie la reţeaua de apă." *Adevărul.Ro.* 2015. https://adevarul.ro/news/bucuresti/foto-s-a-rupt-asfaltulin-zona -tineretului-apa-nova-avarie-1_55a62191f5eaafab2c96b8aa/index.html.

———. "Spectacol printre morminte. Cimitirul Bellu se transformă sâmbătă noapte în cel mai căutat loc din Bucureşti." *Adevărul.Ro.* 2015. https://adevarul.ro/news/bucuresti/spectacol -printre-morminte-cimitirul-bellu-transforma-sambata-noapte-mai-cautat-loc -bucuresri-1_55533ea6cfbe376e35272b72/index.html.

Rainer, Geissmann. "La Claustra." *youtube.com.* 2013. https://www.youtube.com/watch?v =KzUbIu7hcNE.

"Reabiltarea centrului istoric al Bucureştiului." *Igloo.Ro.* Bucharest, 2005. https://www.igloo .ro/articole/reabiltarea-centrului-istoric-al-bucurestiului/.

Reeser, Todd W. and Steven D. Spalding. "Reading Literature/Culture: A Translation of ' as a Cultural Practice.'" *Style* 36, no. 4 (2002): 659–75.

Reilly, Jill. "See out the Apocalypse in Style: Russian Cold War Bunker Is Turned into Luxury Nightclub." *Dailymail.Co.Uk.* December 21, 2012. https://www.dailymail.co.uk/news /article-2250999/Russian-Cold-War-bunker-turned-luxury-nightclub-Tickets-600-hey -you-.html.

Reynolds, Elizabeth. *Underground Urbanism.* New York: Routledge, 2020.

"Rezultatele concursului 'Amenajarea spaţiului supratcran parcaj Universitate: Proiect Tehnic.'" *Arhitectura.* 2012. https://arhitectura-1906.ro/2012/01/rezultatele-concursului-"amenajarea -spatiului-suprateran-parcaj-universitate-proiect-tehnic/

Richards, Jonathan. *Facadism.* New York: Routledge, 1994.

Ritzer, George. *The McDonaldization of Society.* Thousand Oaks, CA: Pine Forge Press Publi- cation, 1996.

"Bucharest, among Europe's Most Polluted Cities." *Romania-Insider.Com.* November 1, 2018.
    https://www.romania-insider.com/bucharest-numbeo-pollution-mid-2018.

Rodgers, Dennis. "'Disembedding' the City: Crime, Insecurity and Spatial Organization in
    Managua, Nicaragua." *Environment and Urbanization* 16, no. 2 (October 1, 2004): 113–24.

Rom Engineering Ltd. and AVENSA Consulting SRL. "Planul de mobilitate urbană durabilă
    2016–2030 regiunea Bucureşti–Ilfov." Bucharest, 2015.

"Romania's Capital Mayor." *The Economist.* London, September 2000. http://www.economist
    .com/node/360036.

"Romanians Flock to Ceauşescu's Palace." *Transitions Online.* 2016.

Rose, Mark and Şengül Aydingün. "Under Istanbul." *Archaeology* 60, no. 4 (2007): 34–40.

Rose, Nikolas. "Governing the Enterprising Self." In *The Values of the Enterprise Culture:
    The Moral Debate,* edited by Paul Heelas and Paul Morris, 141–64. London: Routledge,
    1992.

Rotariu, Victor. "Cum va arăta metroul Bucureştean în 2030." *Ziarul Financiar.* 2007. http://
    www.zf.ro/eveniment/cum-va-arata-metroul-bucurestean-in-2030-3059222.

Rudniţchi, Constantin. "De ce parcarea de la Străuleşti e goală, iar parcările de la mall-uri sunt
    arhipline." *România vocile lumii.* 2019.

Rusu, Mihai Stelian. "The Privatization of Death: The Emergence of Private Cemeteries in Ro-
    mania's Postsocialist Deathscape." *Journal of Southeast European and Black Sea* 20, no. 4
    (2020): 571–91.

Sadana, Rashmi. *The Moving City: Scenes from the Delhi Metro and the Social Life of Infra-
    structure.* Oakland: University of California Press, 2021.

Samoilă, Ionela. "Morţii dintre vii: Cimitire care înghit cartiere, oameni nepăsători şi o lege
    greu de aplicat." *Ziare.Com.* 2012. https://m.ziare.com/social/mortii-dintre-vii-cimitire
    -care-inghit-cartiere-oameni-nepasatori-si-o-lege-greu-de-aplicat-reportaj-ziare-com
    -1299984-font3.

Samuels, Kathryn Lafrenz. "Heritage Development: Culture and Heritage at the World Bank."
    In *The Cultural Turn in International Aid: Impacts and Challenges for Heritage and the
    Creative Industries,* edited by Sophia Labadi, 55–72. London: Routledge, 2019.

Sassen, Saskia. *The Global City: New York, London, Tokyo.* Princeton, NJ: Princeton Univer-
    sity Press, 2001.

Săveanu, Simion. *Enigmele Bucureştilor.* Bucharest: Ed. Pentru Turism, 1973.

Savu, Alin, Ştefan Lipan, and Magdalena Crăciun. "Preparing for a 'Good Life': Extracurricu-
    lars and the Romanian Middle Class." *East European Politics and Societies and Cultures*
    34, no. 2 (2020): 485–504.

Scarry, Elaine. *Thermonuclear Monarchy: Choosing Between Democracy and Doom.* New York:
    W. W. Norton, 2014.

Schaffer, Ronald. *Wings of Judgment: American Bombing in World War II.* Oxford: Oxford Uni-
    versity Press, 1988.

Schofield, John and Luise Rellensmann. "Underground Heritage: Berlin Techno and the
    Changing City." *Heritage & Society* 8, no. 2 (July 3, 2015): 111–38.

Sedgewick, Augustine. *Coffeeland: One Man's Dark Empire and the Making of Our Favorite
    Drug.* New York: Penguin Books, 2021.

Semo, Marc. "Une Ville de Chien." *Libération.* 2001. https://www.liberation.fr/planete/2001/01
    /11/une-ville-de-chien_350600.

Serviciul pentru Relația cu Mass-Media. "Conferința primarului general Adrian Videanu: Planul general de transport al municipiului București." *www1.Pmb.Ro.* 2008. http://www1 .pmb.ro/pmb/primar/cpresa/2008/conferinte/conf_2008-01-31.html.

Sherouse, Perry. "Where the Sidewalk Ends: Automobility and Shame in Tbilisi, Georgia." *Cultural Anthropology* 33, no. 3 (2018): 444–72.

Shillito, Lisa-Marie et al. "The Microstratigraphy of Middens: Capturing Daily Routine in Rubbish at Neolithic Çatalhöyük, Turkey." In *Antiquity* 85, no. 329 (2011): 1024–38.

Shoup, Donald. "The High Cost of Free Parking." *Journal of Planning Education and Research* 17, no. 3 (1997): 3–20.

Simmel, Georg. "Bridge and Door." *Theory, Culture & Society* 11, no. 1 (1994): 5–10.

———. *The Metropolis and Mental Life*, edited by Richard Sennett. New York: Appleton-Century-Crofts, 1969.

———. "The Stranger." In *The Sociology of Georg Simmel*, edited by Kurt H. Wolff, 1–3. New York: The Free Press, 1950.

Simone, AbdouMaliq. *For the City Yet to Come: Changing African Life in Four Cities.* Durham, NC: Duke University Press, 2004.

Sîrbu, Onny. "'Fantoma' cu birouri din inima Bucureștiului. În cazul unui cutremur mare, clădirea se va prăbuși", afirmă părintele Bogdan de la Biserica Armenească." *Magnanews. Ro.* 2016. https://magnanews.ro/2016/01/fantoma-cu-birouri-din-inima-bucurestiului-in -cazul-unui-cutremur-mare-cladirea-se-va-prabusi-afirma-parintele-bogdan-de-la -biserica-armeneasca/.

Smărăndescu, P. "Casa Cotescu." *Arhitectura.* 2 (1916): 52–59.

Smart, Barry. "Digesting the Modern Diet: Gastro-Porn, Fast Food and Panic Eating." In *The Flaneur*, edited by Keith Tester, 158–80. London: Routledge, 1994.

Smith, Neil. *The New Urban Frontier: Gentrification and the Revanchist City.* London: Routledge, 1996.

Soficaru, A. D. et al. "Altered Shapes, Same People: Scaphocephaly in the Early Modern Bucharest." *Homo* 69, no. 4 (2018): 176–87.

Soficaru, Andrei et al. "Date antropologice preliminare privind osemintele umane din necropola isericii Sf. Sava (Piața Universității, București)." *Revista de cercetări arheologice și numismatice.* 1, no. 1 (2015): 229–55.

Solis, Julia. *New York Underground: The Anatomy of a City.* London: Routledge, 2005.

Solomon, Christopher. "Swiss Weigh Future Role of Bunkers in the Alps." *New York Times.* January 19, 2011. https://www.nytimes.com/2011/01/20/world/europe/20swiss.html.

Spiridon, Monica. "Literature and the Symbolic Engineering of the European Self." *European Review* 17, no. 1 (2009): 149–59.

———. "The Fate of a Stereotype: Little Paris." *PMLA* 122, no. 1 (2007): 271–74.

Stănescu, N. "Case vechi românești." *Arhitectura.* 2 (1916): 60–61.

Starosielski, Nicole. *The Undersea Network.* Durham, NC: Duke University Press, 2015.

Stoian, Cristina. "Proprietarii barurilor din centrul vechi al Bucureștiului preferă să ia amendă decât să renunțe la terase." *Ziarul Financiar.* 2010. https://www.zf.ro/print /5785006.

Stoica, Ionel and Virgil Burlă. "Nababii din subteran. Scumpirile de la metrou s-au făcut prin șantaj." *Evz.Ro.* 2013. https://www.ziarelive.ro/stiri/nababii-din-subteran-scumpirile-de-la -metrou-s-au-facut-prin-santaj.html.

Stuckler, David, Lawrence King, and Martin McKee. "Mass Privatisation and the Post-Communist Mortality Crisis: A Cross-National Analysis." *The Lancet* 373, no. 9661 (January 31, 2009): 399–407.

Sudjic, Deyan. *The Edifice Complex: How the Rich and Powerful Shape the World.* New York: Penguin Press, 2005.

Surcel, Vasile. "Necropola de sub piața Universității." *Jurnalul.Ro.* Accessed April 6, 2017. http://jurnalul.ro/print-561180.html.

Suttor, Greg. "Basement Suites: Demand, Supply, Space, and Technology." *Canadian Geographer* 61, no. 4 (2017): 483–92.

Synergetics Corporation et al. "Plan Integrat de Dezvoltare Urbană Zona Centrală București." Bucharest, 2010.

Tanaka, Yuki, Toshiyuki Tanaka, and Marilyn B. Young. *Bombing Civilians: A Twentieth-Century History.* New York: The New Press, 2010.

Tănase, Bogdan Peter, Ioana Manolache, and Paul Filip. *Panteonul național* Vol. 2. Bucharest: Editura Tradiție, 2010.

Taylor, A. R. E. "4 Future-Proof: Bunkered Data Centres and the Selling of Ultra-Secure Cloud Storage." *Journal of the Royal Anthropological Institute* 27 (2021): 76–94.

Teodorescu, Ruxandra Florina and Valentina Constanța Tudor. "Demographic Analysis of the Bucharest-Ilfov Region." *Scientific Papers Series Management, Economic Engineering in Agriculture and Rural Development* 19, no. 3 (2019): 593–98.

Timu, Andra and Irina Vîlcu. "Facelift for EU's Worst Infrastructure Eases Hit from Virus." *Bloomberg.Com.* October 5, 2020. https://www.bloomberg.com/news/articles/2020-10-06/facelift-for-eu-s-worst-infrastructure-eases-hit-from-virus.

———. "Romanian GDP Growth Quickens More Than Expected After Tax Cuts." *Bloomberg.Com.* May 13, 2016. http://www.bloomberg.com/news/articles/2016-05-13/romanian-gdp-growth-quickens-more-than-expected-after-tax-cuts.

Tiron, Mirabela. "Cea mai vândută carte din România: 22.000 de exemplare. Bestseller-ul din Franța se vinde de 80 de ori mai bine." *Ziarul Financiar.* 2012. https://www.zf.ro/special/cea-mai-vanduta-carte-din-romania-22-000-de-exemplare-bestsellerul-din-franta-se-vinde-de-80-de-ori-mai-bine-9440284.

Toea, Diana. "Un loc de veci la Bellu costă cât jumătate de garsonieră: 10.000 de euro pentru odihnă de veci." *Adevărul.Ro.* 2012. http://adevarul.ro/news/bucuresti/un-loc-veci-bellu-costa-jumatate-garsoniera-10000-euro-odihna-veci-1_50bded8a7c42d5a663d041ab/index.html.

Toma, Laura. "Room without a View: Construction in Bucharest after 1989." *EU Observer.* July 10, 2010.

Toma, V. Ciugudean et al. "Necessary Geological and Geotechnical Information for a Metro Project in an Historical and Urbanised City Area. The Case of 'Metro Bucharest, Line 4 Extension.'" In *Tunnels and Underground Cities: Engineering and Innovation Meet Archaeology, Architecture and Art: Volume 3: Geological and Geotechnical Knowledge and Requirements for Project Implementation,* edited by Daniele Peila, Giulia Viggiani, and Tarcisio Celestino, 711–20. London: CRC Press, 2020.

Tomlinson, John. *Globalization and Culture.* Chicago: University of Chicago Press, 1999.

TomTom International BV. "TomTom Traffic Index: Measuring Congestion Worldwide." *tomtom.com.* 2017.

"Top 10 Tallest Office Buildings in Bucharest." *Romania-Insider.Com*. October 29, 2015. https://www.romania-insider.com/top-10-tallest-office-buildings-in-bucharest.

Topham, Gwyn. "Final Piece in £700m Overhaul of Bank Tube Station in London Opens to Public." *Guardian*. February 27, 2023. https://www.theguardian.com/uk-news/2023/feb/27/final-piece-in-700m-overhaul-of-bank-tube-station-in-london-opens-to-public.

Toth, Jennifer. *The Mole People: Life in the Tunnels Beneath New York City*. Chicago: Chicago Review Press, 1995.

Tousignant, Isa. "Guide to the Underground City." *mtl.org*. 2022. https://www.mtl.org/en/experience/guide-underground-city-shopping.

Trajanescu, Ion D. "Casa din strada Labirint." *Arhitectura*. 2 (1916): 62–64.

Truitt, Allison J. "On the Back of a Motorbike: Middle-Class Mobility in Ho Chi Minh City, Vietnam." *American Ethnologist* 35, no. 1 (2008): 3–19.

Truxal, Luke. "Bombing the Romanian Rail Network." *Air Power History* 65, no. 1 (2018): 15–22.

Tucker, Jonathan. *War of Nerves: Chemical Warfare from World War I to Al-Qaeda*. New York: Knopf Doubleday Publishing Group, 2007.

Turner, Victor W. "Liminality and Communitas." In his *The Ritual Process: Structure and Anti-Structure*, 94–130. New York: Routledge, 2017.

Turnock, David. "Housing Policy in Romania." In *Housing Policies in Eastern Europe and the Soviet Union*, edited by J. A. A. Sillince, 134–69. New York: Routledge, 1990.

———. "The Planning of Rural Settlement in Romania." *The Geographical Journal* 157, no. 3 (November 1, 1991): 251–64.

Turp-Balazs, Craig. "Bucharest's Spring Palace: Where Kitsch Knows No Bounds." *Emerging Europe*. July 4, 2020. https://emerging-europe.com/after-hours/bucharests-spring-palace-where-kitsch-knows-no-bounds/.

"Un buncăr subteran care asigură atât supraviețuirea cât și confortul." *www.Bihor.Ro*. 2014. https://www.bihon.ro/stirile-judetului-bihor/un-buncar-subteran-care-asigura-atat-supravietuirea-cat-si-confortul-p-299656/.

"Un lac din București a început să sece după ridicarea unui bloc-țeapă." *Green-Report.Ro*. 2009. https://www.green-report.ro/gandul-un-lac-din-bucuresti-inceput-sa-sece-dupa-ridicarea-unui-bloc-teapa/.

"Un loc de veci la cimitirul Bellu a ajuns să coste peste 10.000 de euro, cât jumătate de garsonieră." *Stirileprotv.Ro*. 2012. https://stirileprotv.ro/stiri/actualitate/un-loc-de-veci-la-cimitirul-bellu-a-ajuns-sa-coste-peste-10-000-de-euro-cat-jumatate-de-garsoniera.html.

Vähäaho, Ilkka. "Underground Space Planning in Helsinki." *Journal of Rock Mechanics and Geotechnical Engineering* 6, no. 5 (2014): 387–98.

Varga, Vasile. *Nicolae Grigorescu*. Iași: Meridiane, 1973.

Venkatesh, Sudhir Alladi. *Off the Books: The Underground Economy of the Urban Poor*. Cambridge, MA: Harvard University Press, 2009.

Verdery, Katherine. *The Political Lives of Dead Bodies: Reburial and Postsocialist Change*. New York: Columbia University Press, 1999.

———. *The Vanishing Hectare: Property and Value in Postsocialist Transylvania*. Ithaca, NY: Cornell University Press, 2003.

———. *What Was Socialism, and What Comes Next?* Princeton, NJ: Princeton University Press, 1996.

Verona, Roxana. "Bucharest at the crossroads." *PMLA* 122, no. 1 (2007): 275–80.

Vice Staff. "The VICE guide to Bucharest 2014." *Vice.Com*. 2014. https://www.vice.com/en_us /article/5gkwwa/the-vice-guide-to-bucharest-2014-345.

Vigroux, Grégoire. "What Makes Romania a Preferred Outsourcing Destination?" *Telus International*. 2015. http://telusinternational-europe.com/romania-preferred-outsourcing -destination/.

Vintilă, Petru. *Bucureştiul subpământean / Underground Bucharest*. Bucharest: Heliopolis, 2006.

Virilio, Paul. *The Original Accident*. London: Wiley, 2007.

VisitSeoul.net. "Seoul Underground—A Different Plane of Time." *VisitSeoul.net*. 2020. https:// english.visitseoul.net/walking-tour/Seoul-Underground/34964?letterSn =1822&ltcontent=220521_Seoul Underground%3Cbr%3EA Different Plane of Time.

Voinea, Mihai, and Cristian Delcea. "Interviu Povestea venirii McDonald's în România: '16.000 de oameni au spart geamurile la deschidere.'" *Adevărul.Ro*. August 2015.

Wacquant, Loïc. "Urban Outcasts: Stigma and Division in the Black American Ghetto and the French Urban Periphery." *International Journal of Urban and Regional Research* 17, no. 3 (1993): 366–83.

———. "Making Class: The Middle Class(es) in Social Theory and Social Structure." In *Bringing Class Back In: Contemporary and Historical Perspectives*, edited by Scott Mcnall, 39– 64. Boulder, CO: Westview Press, 1991.

Wainwright, Oliver. "Billionaires' Basements: The Luxury Bunkers Making Holes in London Streets." *Guardian*. November 9, 2009. https://www.theguardian.com/artanddesign/2012 /nov/09/billionaires-basements-london-houses-architecture.

———. "Is It the End for Millionaire Mega-Basements?" *Guardian*. October 15, 2013. https:// www.theguardian.com/artanddesign/2013/oct/15/mega-basement-extensions-planning -policy.

Warner, Michael. "Publics and Counterpublics." *Public Culture* 14, no. 1 (2002): 49–90.

Webster, George S. "Subterranean Street Planning." *The Annals of the American Academy of Political and Social Science* 51, no. 1 (1914): 200–207.

Williams, Raymond. *The Country and the City*. New York: Oxford University Press , 1975.

Williams, Rosalind H. *Notes on the Underground: An Essay on Technology, Society, and the Imagination*. Cambridge, MA: MIT Press, 2008.

Wilmott, Clancy. "Surface: Seeing, Solidifying, and Scaling Urban Space in Hong Kong." In *Voluminous States: Sovereignty, Materiality, and the Territorial Imagination*, edited by Franck Billé, 146–58. Durham, NC: Duke University Press, 2020.

Wilson, Louise K. "Sounds from the Bunker: Aural Culture and the Remainder of the Cold War." *Journal of War and Culture Studies* 13, no. 1 (2020): 33–53.

Wilson, William Julius. *The Truly Disadvantaged: The Inner City, the Underclass, and Public Policy*. Chicago: University of Chicago Press, 1990.

Woodhouse, Skylar. "What Penn Station's $6 Billion Makeover Means for NYC." *Bloomberg .Com*. August 18, 2022. https://www.bloomberg.com/news/articles/2022-08-18/what-penn -station-s-6-billion-makeover-means-for-nyc-quicktake.

World Economic Forum. "Quality of Roads." Geneva: World Economic Forum, 2016. https:// reports.weforum.org/pdf/gci-2017-2018-scorecard/WEF_GCI_2017_2018_Scorecard _EOSQ057.pdf.

WSP Global. "Taking Urban Development Underground: Future-Ready Solutions for Ensuring Urban Sustainability." Montreal, 2021.

Yeh, Rihan. *Passing: Two Publics in a Mexican Border City.* Chicago: University of Chicago Press, 2017.

Yoshida, N. and S. Nakamura. "Damage to Daikai Subway Station during the 1995 Hyogoken-Nunbu Earthquake and Its Investigation." In *Eleventh World Conference on Earthquake Engineering,* 283–300. Amsterdam: Elsevier Science, 1996.

Young, Iris Marion. *Justice and the Politics of Difference.* Princeton, NJ: Princeton University Press, 1990.

Yurchak, Alexei. "Russian Neoliberal: The Entrepreneurial Ethic and the Spirit of 'True Careerism.'" *Russian Review* 62, no. 1 (2003): 72–90.

Zachmann, Sebastian. "România e pe ultimul loc în Europa în privința pieței de carte." *Adevărul.Ro.* 2018. https://adevarul.ro/news/politica/ovidiuraetchi-romania-e-ultimul-loc -europa-privintapietei-carte-rusine-anul-centenarului—8-masuri-dublarea-consumu lui-carte-1_5aae4e1edf52022f75b79c35/index.html.

Zhuang, Haiyang et al. "Seismic Performance Levels of a Large Underground Subway Station in Different Soil Foundations." *Journal of Earthquake Engineering* 25, no. 14 (2019): 1–26.

Ziare.com. "#Colectiv: Tavanul s-a aprins ca o baltă de benzină—Mărturiile unui supra viețuitor.." *Ziare.Com.* 2015. https://ziare.com/stiri/incendiu-club-colectiv/colectiv-tavanul -s-a-aprins-ca-o-balta-de-benzina-marturiile-unui-supravietuitor-1397239.

———. "Românii au bani de bere, dar nu și de cărți." *Ziare.Com.* August 13, 2013. http://www .ziare.com/social/romani/romanii-au-bani-de-bere-dar-nu-si-de-carti-1251230.

Ziua News. "Directorul General Metrorex și Virgil Măgureanu, presiuni și lobby la ministerul transporturilor pentru privilegiile sindicalistului Ion Rădoi." *ZiuaNews.Ro.* 2016. https:// www.ziuanews.ro/dezvaluiri-investigatii/directorul-general-metrorex-i-virgil-m-gure anu-presiuni-i-lobby-la-ministerul-transporturilor-pentru-privilegiile-sindicalistului -ion-r-doi-597588.

Zukin, Sharon. "Gentrification: Culture and Capital in the Urban Core." *Annual Review of Sociology* 13 (1987): 129–47.

INDEX

ACKNOWLEDGMENTS

For this book to come together, a lot of overscheduled people had to offer me their time, trust me with their stories, and open up their contacts to me. I am deeply indebted to each and every person who took my calls, put up with my questions, entrusted me with access, offered me their perspective, and introduced me to their colleagues and friends.

The preliminary research for this book, which occurred between June 2010 and November 2011, was made possible by the support of the Fulbright-Hays program, the National Science Foundation's DDRA, the Student Fulbright Program, and funds from Stanford University. The primary research was supported by a Wenner-Gren Foundation Post-Ph.D. Research Grant, Mellon Faculty Grants, and by the support of the Office of the Vice President for Research at Saint Louis University. In addition to travel, these funds allowed me to hire very smart and expertly trained research assistants from the University of Bucharest. My deepest thanks go to Anca Yastremskyi and to Mara Folcic who helped to arrange meetings and to carefully transcribe recorded interviews. They also helped me to think through the many twists and turns of research to find new paths forward when confronting apparent dead ends.

Writing articles helped me to work through the book's empirical and analytical commitments. Aspects of this research appeared as articles in the journals *Environment and Planning A* 52, no. 7 (2020), as "Segmenting the City"; *Journal of the Royal Anthropological Institute* 28, no. 1 (2022), as "Up, Down, and Away"; *Anthropological Quarterly* 95, no. 4 (2022), as "The Digital Underground"; and *Current Anthropology* 64, no. 1 (2023), as "The Subject of the Underground." I am grateful to the editors and anonymous reviewers of these articles for their insights, criticisms, and affirmations. Importantly, for this book, I have reworked, expanded, and in some instances completely rethought the ethnographic material that appeared in these articles. I am also grateful for the invaluable feedback that I received from conversations during invited talks and workshops at the University of Sheffield's Urban Institute ("Volumetric Urbanism International Workshop"), the University of Pennsylvania ("Social Impacts of Post-Socialist Transition and Policies for the Future"), Washington University in St. Louis ("Ethnographic Theory Workshop"), and Indiana University (keynote lecture for the Twelfth Annual International Romanian Studies Conference), as well as from conference panels at the Annual Meetings of the American Anthropological Association, the Annual American Ethnological Society Conference, and the Central Slavic Conference Annual Meetings.

I am especially indebted to Liviu Chelcea, Bruce Grant, David Pike, Robert Samet, and Rihan Yeh for their collective efforts to workshop an advanced draft of this manuscript. Their keen insights and suggestions came at a critical moment and helped to ready the manuscript for review. My brother, Kevin, who is also an anthropologist, kindly moderated the workshop. He also offered his encouragements and enthusiasm throughout when I needed it most. Philip Sayers provided editorial guidance as I revised the manuscript for submission.

I am also thankful to those who engaged my thoughts through conversations and questions, whether in passing or over the years: Nikhil Anand, Rowland Atkinson, Andrea Ballestero, Luke Bennett, Pete Benson, Sarah Besky, John Bowen, Maria Bucur, Liz Chiarello, Amy Cooper, Talia Dan-Cohen, Martin Dodge, Dace Dzenovska, Terra Edwards, Maria Escallon, Dan Falcan, Bradley Garrett, Kristen Ghodsee, Daniela Giudici, Stephen Graham, Bruce Grant, Maron Greenleaf, Andreas Hackl, Scott Harris, Stefan Hohne, Helen Human, Gail Kligman, Alejandra Leal, Anru Lee, Alaina Lemon, Daniel Mains, Simon Marvin, Ramah McKay, Donald McNeill, Caroline Melly, Madalina Minu, Alex Nading, Mitchell Orenstein, Federico Perez, Maria Alejandra Perez, Claudiu Popa, Chris Prenner, Ato Quayson, Laurence Ralph, Dennis Rodgers, Anca Roncea, Rashmi Sadana, Jonathan Silver, Harris Solomon, Claudio Sopranzetti, and Kedron Thomas. Thank you all.

I first conceptualized this book project while completing my dissertation at Stanford University. My deepest thanks go to Jim Ferguson, Liisa Malkki, and Lissa Caldwell for their direction and support at this book's earliest moments. At Saint Louis University, I have been fortunate to enjoy an interdisciplinary and incredibly supportive home in the Department of

Sociology and Anthropology where this project could take full shape. My special thanks go to my department chairs, first Ric Colignon and then Joel Jennings, for helping me to secure the time and resources needed to complete this book.

I owe a very large debt of gratitude to everyone at the University of Pennsylvania Press. My thanks go to the editors of this book series, "The City in the Twenty-First Century," Genie Birch and Susan Wachter, for their excitement and commitment to this book upon receiving the proposal. I am also deeply indebted to my editor, Robert Lockhart, for his unwavering support and for his steady editorial guidance while moving the book through production.

Of course, absolutely none of this would have been possible without the loving support of my family, which grew alongside this project. My daughter Rosemary was born as the principal fieldwork for Underground got under way. My wife Helen, my in-laws Kate and Dan, and my parents Bruce and Mary absorbed the many responsibilities and demands of my growing household so that I could travel to the field. Even after picking up the slack for my share of diaper changes, dinner preps, and daycare drop-offs while I was away, they warmly welcomed me back from my final research trip in December of 2019. With my son on the way, we collectively celebrated having put the difficulties of fieldwork with young children in the rearview mirror. We happily looked forward to what we expected would be a quiet writing sabbatical ahead. A few months later, as I sketched the most provisional of outlines for this book, the COVID-19 virus upended life the world over. Sitting at home in lockdown, with daycare shuttered, a newborn imminent, and with the dawning realization that an end to the pandemic was nowhere in sight, I vividly remember worrying to Helen that the time to write this book would be lost to the practical realities of pandemic parenting. My family then came together to give me their unflinching support. "Bubbled" with my in-laws, Helen, Kate, and Dan created for me three uninterrupted hours of writing time each day so that I could draft this book. For that I am forever grateful.

During that same period, nothing motivated me more to write this book than my desire to dedicate it to my kids. Each day, Rosemary played just quietly enough, while Leo slept just long enough and just consistently enough, for me to write. Rosemary also dutifully oversaw a calendar that tracked my progress. When I met my writing deadlines, Rosemary marked these tiny victories by hosting a "cupcake party." Not wanting to disappoint her expectations, I made sure to never miss a deadline and to always remain on schedule. In this sense, my kids pressed me to make timely decisions and to stick to them, to not procrastinate, and to keep pushing the analysis forward one paragraph at a time, one day at a time, even as the world around us unraveled. That is why I dedicate this book with all my love to Rosemary and Leo.

www.ingramcontent.com/pod-product-compliance
Lightning Source LLC
Chambersburg PA
CBHW032346280326
41935CB00008B/470